Mobile Robotics with Arduino
Design and Programming

Klaus Röbenack

Copyright © 2018 Klaus Röbenack
All rights reserved.

Röbenack, Klaus:
Mobile Robotics with Arduino
Design and Programming.

ISBN-13:   978-1726432337
ISBN-10:   1726432335

All rights reserved. No part of this book may be reproduced, in any form or by any means, without the permission of the publisher.

Prof. Dr. Klaus Röbenack
Brucknerstr. 17, D-01309 Dresden, Germany

Printed by CreateSpace, An Amazon.com Company

The author and the publisher of this book have used best efforts in preparing this book. Neither the author nor the publisher guarantee the accuracy or completeness of any information published herein. The author and the publisher make no warranty of any kind, expressed or implied, with regard to the information contained in this book. Neither the author nor the publisher shall be responsible for any errors, omissions, or claims for damages, including exemplary damages, arising out of use, inability to use, or with regard to the accuracy or sufficiency of the information contained herein.

All product names mentioned in this book are the trademarks of their respective owners. Arduino® brands and logos are trademarks of Arduino AG.

Preface

The robot platform described in the book was developed in connection with the school project week of the Martin-Andersen-Nexö-Gymnasium (MANOS) in Dresden, Germany. For several years, the Institute of Control Theory at the Technische Universität Dresden has been organizing school pupil internships on tracking control of mobile robots as part of this project week.

The book describes the design and programming of mobile robots. The Arduino platform, which is easy to use, was chosen to control the robot. The author describes the interfacing and programming of typical components such as motors, LCD modules and various sensors up to the operation of an infrared remote control or a radio remote control. In contrast to ready-to-use robot kits, the reader is also given the necessary freedom to implement his own ideas.

This book is intended for readers who already have some experience with microcontrollers in general or the Arduino platform in particular. In addition, a basic knowledge of electronics and the ability to create simple programs in C or C++ are expected.

I would like to thank Dr. Carsten Knoll and Dipl.-Ing. Christian John for their commitment during the student project weeks. My thanks also go to Dr. Jan Winkler, who supported me in many ways. I would also like to thank Anja Lehmann, Matthias Schäfer and Mirko Franke for their efforts in carrying out the student internships. In addition, Mr. Schäfer deserves thanks for the review of the manuscript and his suggestions. Furthermore, I would like to thank Prof. Dr. Frank Woittennek, Oliver Schnabel and Marcus Riesmeier for further developments of the robot platform.

This book was first published in the German language under the title "Mobiler Eigenbauroboter mit Arduino: Aufbau und Programmierung". Besides the translation, I also adapted the references such that they now refer to contributions in the English language.

I want to dedicate this book to my children. Last but not least, I would like to thank my wife for her daily support.

Dresden, September 2018 — Klaus Röbenack

Contents

Preface		iii
1 Arduino Platform		**1**
1.1	Introduction	1
1.2	Arduino Boards	1
	1.2.1 Arduino Uno	1
	1.2.2 Arduino Leonardo	2
	1.2.3 Alternative Boards	3
1.3	Integrated Development Environment	5
1.4	Programming Language	6
	References	8
2 Notes on the Construction of the Robot		**9**
2.1	Selection of the Motors	9
2.2	Batteries	11
2.3	Mechanical Design	12
	References	16
3 Motor Control		**17**
3.1	Speed Control with Pulse Width Modulation (PWM)	17
3.2	Arduino Motor Shield R3	20
3.3	Velleman Motor Shield	24
3.4	C++ Class for Motor Control	27
3.5	Motor Control Library	31
	References	35

4 LCD Output — 37
- 4.1 Connecting an LCD Module — 37
- 4.2 Control of the LCD Module — 40
- 4.3 Formatting the Output — 41
- References — 44

5 Buttons and Switches — 45
- 5.1 Connecting Push-Buttons — 45
- 5.2 Obstacle Avoidance using Push-Buttons — 49
- 5.3 Wired Remote Control with Switches — 52
- 5.4 Additional Digital Inputs and Outputs on Arduino Leonardo — 57
- References — 60

6 Measurement of Analog Signals — 61
- 6.1 Connecting a Potentiometer — 61
- 6.2 Wired Control via two Sliding Potentiometers — 65
- 6.3 Current Measurement on the Arduino Motor Shield — 67
- 6.4 Obstacle Avoidance by Means of Motor Current Measurement — 70
- 6.5 Motor Voltage Measurement on the Velleman Motor Shield — 73
- References — 74

7 Distance Measurement — 75
- 7.1 Distance Sensors based on Optical Triangulation — 75
- 7.2 Calibration of the Sensor — 79
- 7.3 Obstacle Avoidance with one Distance Sensor — 82
- 7.4 Capturing the Surroundings and Searching for a new Direction — 84
- 7.5 Obstacle Avoidance with two Distance Sensors — 90
- 7.6 Distance Measurement using Ultrasonic Sensors — 94
  - 7.6.1 Ultrasonic Distance Measurement Principle — 94
  - 7.6.2 Operation of the Ultrasonic Distance Sensors HC-SR04 and HY-SRF05 — 95
  - 7.6.3 Single-wire Operation of the Ultrasonic Distance Sensor HC-SR04 — 100
- References — 103

8 Line Detection and Tracking — **105**
- 8.1 Line Detection with a Reflective Optical Sensor 105
- 8.2 Line Tracking with one Reflective Optical Sensor 108
- 8.3 Line Tracking with Several Reflective Optical Sensors 110
- References 112

9 Wireless Control of the Robot — **113**
- 9.1 Infrared Remote Control 113
  - 9.1.1 Connecting the Receiver Module and Requesting the IR Codes 113
  - 9.1.2 Motor Control via IR Codes 116
- 9.2 Radio Remote Control 118
  - 9.2.1 Connecting the Receiver 118
  - 9.2.2 Conversion of the Signals for Motor Control 121
- References 125

10 Additional Design Variants — **127**
- 10.1 LCD KeyPad Shield 127
- 10.2 Motor Driver with L298N 131
- 10.3 Obstacle Detection with IR Sensors 134
- References 136

Chapter 1

Arduino Platform

1.1 Introduction

Arduino is a microcontroller platform that includes both hardware and open source software [1, 6]. In addition, the hardware is open source in the sense that the circuit diagrams, the PCB layout, etc. are available online. There are several books [5] and starter kits for the Arduino platform.

Numerous different boards are available within the Arduino platform. The boards house the microcontroller. In most cases it is a controller of the ATmega series from Atmel [7]. Several analog inputs as well as digital inputs and outputs are provided by the microcontroller. In addition to the Arduino controller boards, there are many extension boards called *shields*.

1.2 Arduino Boards

This section briefly introduces the Arduino boards intended for the robot.

1.2.1 Arduino Uno

The Arduino Uno was probably the most popular Arduino board at the time of writing. The board is produced with the controller ATmega328. This controller has 32 KiB flash memory, which can mainly be used for our own programs. The clock frequency is generated by a 16 MHz quartz. Communication with the PC runs via a USB interface, which is controlled by an ATmega8U2 or from revision 3 by an ATmega16U2. The board also has 14 digital input/output pins with the numbers 0,...,13 and 6 analog input pins with the designations A0,...,A5 (see Fig. 1.1). In addition, during revision 3 one of the connectors was extended by two pins for the signals SDA and SCL. With these pins you can operate a serial data bus communicating via two signal lines, known as I^2C or TWI (two-wire interface) [8]. While the ATmega328 microcontroller was originally installed in the DIP28 package variant, now almost exclusively the SMD variants used.

Figure 1.1: Arduino Uno R3, SMD variant

1.2.2 Arduino Leonardo

The special feature of Arduino Leonardo is the use of a microcontroller with integrated USB interface. This not only saves the USB to UART bridge, but allows the Leonardo board to emulate a keyboard or a mouse w.r.t. the PC. It is also possible to use the analog inputs as digital inputs or outputs. If this option is used, a total number of 20 digital inputs or outputs are available. The analog inputs A0,...,A5 are then assigned to the digital channels 18,...,23 (see also Section 5.4 from p. 57). Conversely, six of the digital channels can also be used as additional analog inputs A6,...,A11, so that a total number of 12 analog inputs can be used. These additional analog channels are indicated on the conductor side of the board. The component side of the board can be seen in Fig. 1.2.

Figure 1.2: Arduino Leonardo

1.2.3 Alternative Boards

There are numerous alternatives to the standard boards described. This section is intended to provide only a few suggestions and therefore does not claim to be exhaustive.

Boards from other manufacturers: Some manufacturers offer alternative boards that are compatible with Arduino, e.g. the Seeeduino Board from Seeed Technology [3] or the RedBoard from SparkFun Electronics [4].

Boards with more ports: The 14 digital and 6 analog pins of the standard boards are quickly occupied. Boards like Arduino Mega or Arduino Mega 2560 with 54 digital and 16 analog pins have considerably more inputs and outputs (see Fig. 1.3). The ATmega1280 and ATmega2560 controllers are used for these boards.[1]

---

[1] Alternatively, Arduino Due can be used, which has a 32 bit controller with ARM core. Like the two Arduino Mega Boards, this board has 54 digital channels, but only 12 analog inputs. However, Arduino Due operates at 3.3 V (instead of the usual voltage of 5 V). This means that sensors and shields that output 5 V cannot be connected directly to the board. In these cases, level adjustments would be necessary.

Smaller boards: For a space-saving construction there are also several boards with a smaller form factor, for example Arduino Micro, Arduino Mini or Arduino Nano. Arduino Nano corresponds in functionality to Arduino Uno, Arduino Micro shown in Fig. 1.4 corresponds to Arduino Leonardo. Smaller boards are also offered by other manufacturers, e.g. Arduino Pro Mini and Pro Micro from SparkFun Electronics [4].

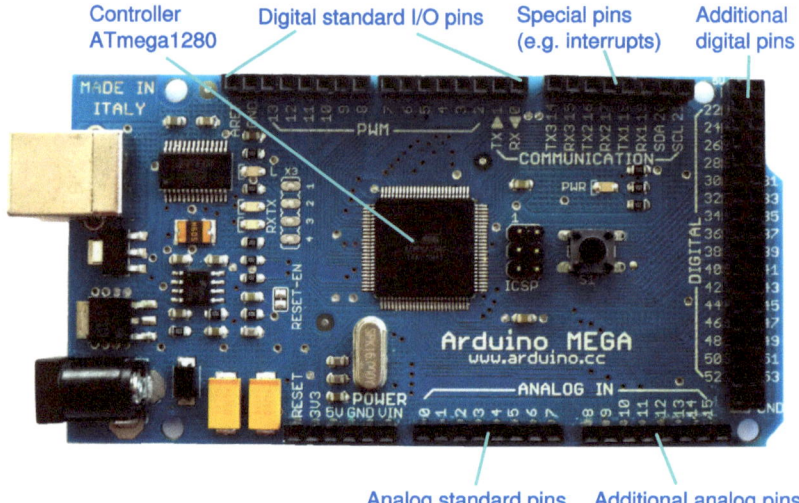

Figure 1.3: Arduino Mega

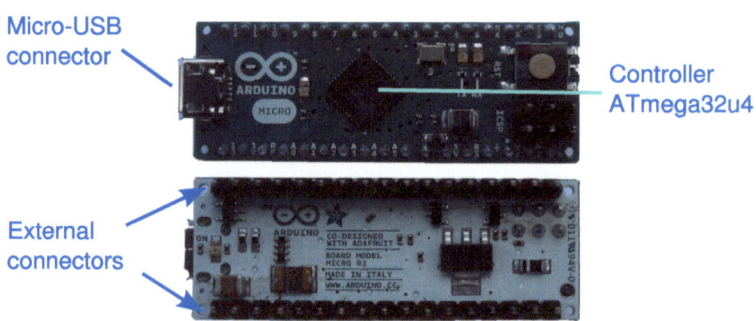

Figure 1.4: Arduino Micro (both sides)

1.3 Integrated Development Environment

Arduino programs can be written online using the Arduino Web Editor [2]. This is the recommended way to develop Arduino programs provided a reliable internet connection is available.

Alternatively, an open source integrated development environment (IDE) can be downloaded from the Arduino website [1]. The software is available for common PC operating systems. The installation is well documented on the Arduino website. Depending on the Arduino board used, the installation of additional drivers for the USB interface may be necessary. Under Linux, the installation can be performed using the common package managers (e.g. `aptitude`, `KPackage`, `PackageKit`, `YaST`, `yum`, `yumex`).

The Arduino IDE essentially consists of an editor for the source code and functions for compiling the source code and uploading it to an Arduino board. Additionally it is possible to display messages of the Arduino board on the PC. In earlier versions the source files had the extension `.pde`, in newer versions `.ino`.

To start up, the Arduino board should first be connected to the PC via a USB cable. Next select on the Arduino IDE under Tools > Board the used Arduino board. Under Tools > Serial Port you can choose the USB port used for the communication with the Arduino board. Depending on the PC operating system and the Arduino board, several channels are available.

Numerous examples for common programming tasks, especially for controlling the peripherals, are available on the Arduino IDE under File > Examples. For an initial test, the `Blink` program can be called up under File > Examples > 01.Basics > Blink, which makes the LED (connected to pin 13 on almost all boards) blink. This program must first be compiled (Ctrl+R) and then uploaded to the board (Ctrl+U), see Fig. 1.5.

It is also possible to send messages from the Arduino board back to the PC and display them there. This data transfer takes place via the USB interface. The Arduino IDE has a suitable display program which can be called via Tools > Serial Monitor.

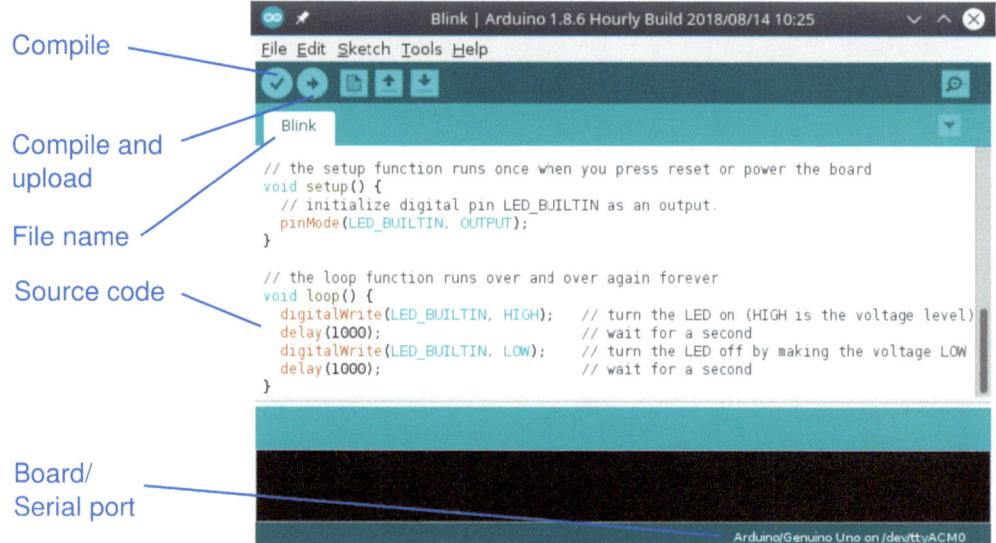

Figure 1.5: Arduino IDE with example program `Blink`

1.4 Programming Language

Arduino is essentially programmed in a C/C++ variant. A C program usually consists of several functions. With the standard variants of C or C++ for PCs or workstations, the function `main` is called with the program start. When the function `main` ends or is processed, the program is also terminated.

The C/C++ variant for Arduino has two special functions. The function `setup` is called once at program start. Therefore, the function `setup` can be used for initialization (e.g. of the input and output channels). After the function `setup` has been executed, the function `loop` is called in an endless loop. The code contained in it is thus cyclically processed again and again (see Fig. 1.6).

1.4 Programming Language

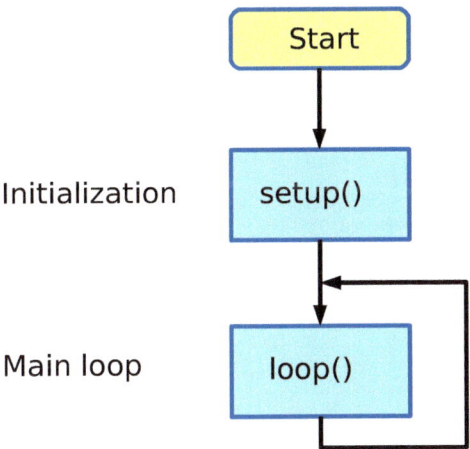

Figure 1.6: Basic structure of an Arduino program

A template for an Arduino program can be found under File > Examples > 01.Basics > BareMinimum:

```
void setup() {
  // put your setup code here, to run once:

}

void loop() {
  // put your main code here, to run repeatedly:

}
```

In the Arduino variant of the C/C++ programming language there are several additional functions that support the hardware of the microcontroller. This applies, for example, to setting the data direction for the digital channels (input or output) or reading data from the analog and digital inputs. The corresponding functions will be introduced in the book.

References

[1] *Arduino — Home.* https://arduino.cc.

[2] *Arduino Web Editor.* https://create.arduino.cc/editor.

[3] *Seeed Technology Co., Ltd.* https://www.seeedstudio.com.

[4] *SparkFun Electronics.* https://www.sparkfun.com/.

[5] Boxall, John: *Arduino Workshop: A Hands-On Introduction with 65 Projects.* No Starch Press, 2013.

[6] Wikipedia contributors: *Arduino — Wikipedia, the free encyclopedia.* https://en.wikipedia.org/w/index.php?title=Arduino.

[7] Wikipedia contributors: *AVR microcontrollers — Wikipedia, the free encyclopedia.* https://en.wikipedia.org/w/index.php?title=AVR_microcontrollers.

[8] Wikipedia contributors: I^2C — *Wikipedia, the free encyclopedia.* https://en.wikipedia.org/w/index.php?title=I%C2%B2C.

Chapter 2

Notes on the Construction of the Robot

2.1 Selection of the Motors

In mobile robotics, permanent magnet DC motors are used almost exclusively. DC motors are typically designed for comparatively high speeds, which is why a gear reduction is also required. For this reason, so-called *gear motors* are preferably used, which combine motor and gearbox in one drive module.

A gear motor used in mobile robotics is sold by Igarashi Motors Ltd. under the type designation 20GN152025-330-050 (see Fig. 2.1). This motor is designed for $4\ldots 12$ V. The gearbox has a transmission ratio of $1:50$. Similar motors are also available from Igarashi for other voltages or transmission ratios. Larger motors can be found in the Modelcraft RB-35 range of gear motors. Here too, the motors are available for different nominal voltages or transmission ratios. The model shown is also designed for a nominal voltage of 12 V.

An alternative to expensive gear motors is the conversion of standard servos [3]. Servos are drives with integrated position control. In addition to a DC motor and gearbox, they also include control electronics. In normal use, servos have only a finite actuating angle, typically in the range of $\pm 90°$. To convert a servo into a gear motor, the control electronics must be removed or deactivated. The connections of the DC motor must then be routed directly to the outside. In addition, the mechanical stops limiting the movement of the

Figure 2.1: 12V gear motors: motor from Igarashi (left), RB-35 gear motor from Modelcraft (right)

servo must be removed. This concerns the potentiometer no longer required for angle measurement as well as possible stops on gears or the axis. Besides the low price, the use of servos is also advantageous in that suitable plastic wheels are offered (Fig. 2.2).

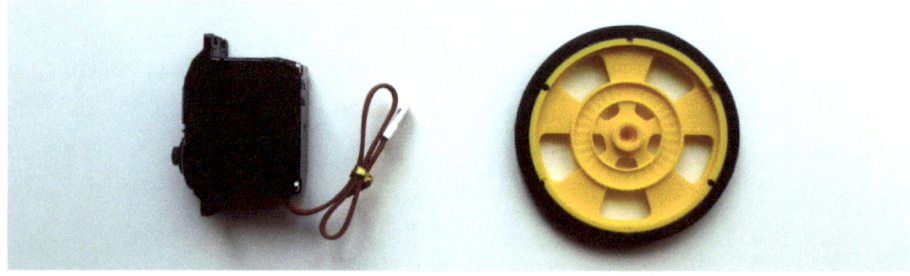

Figure 2.2: Standard servo from Modelcraft with plastic wheel from Solarbotics [2]

Servos are normally designed for a voltage range of $4, 5 \ldots 6$ V. Accordingly, the gear motors resulting from modified servos should not be operated at too high voltages. An operation up to 9 V should not be a problem. However, there are also so-called high-voltage servos, which are usually much more expensive. Manual servo conversion can be avoided by using special servos from the company Solarbotics [2] (Fig. 2.3).

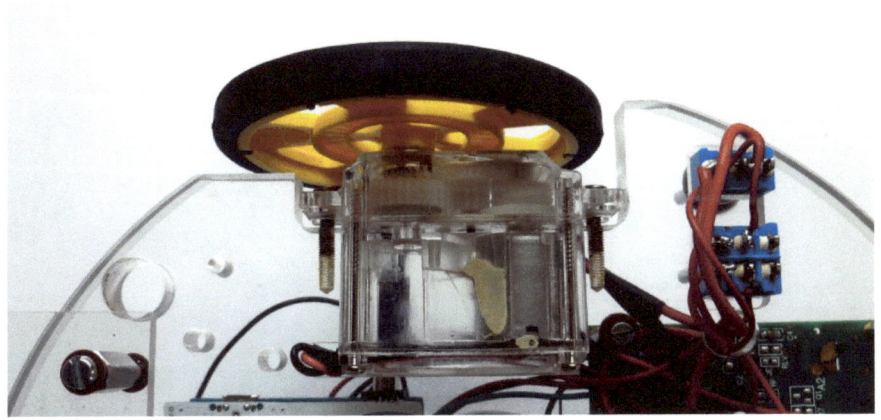

Figure 2.3: Special servo from Solarbotics with a plastic wheel built into the robot

2.2 Batteries

The Arduino boards under consideration require a supply voltage in the range of $6\ldots20\,\text{V}$. An operation with more than $12\,\text{V}$ is not really recommended. A maximum voltage of $12\,\text{V}$ is also specified for the Arduino motor shield described later. We therefore limit the supply voltage range to $7\ldots12\,\text{V}$.

If the mobile robot is frequently used, the use of primary cells makes neither financial nor ecological sense. Therefore, rechargeable batteries (accumulators) will be used. Two different types of batteries are used in model building, namely nickel-metal hydride (NiMH) and lithium-polymer (LiPo) batteries.

A single NiMH cell has a nominal voltage of $1.2\,\text{V}$, immediately after charging the voltage will be at $1.4\ldots1.5\,\text{V}$. For the desired voltage range you can either use several individual batteries or a battery pack. A 6 cell battery pack would provide a nominal voltage of $6\times1.2\,\text{V} = 7.2\,\text{V}$, a 8 cell battery pack would provide $8\times1.2\,\text{V} = 9.6\,\text{V}$. Both variants would be suitable for use with mobile robots. For the connection of battery packs Tamiya connectors are common (see Fig. 2.4). However, a charger with a compatible connector is also required.

Instead of NiMH batteries, LiPo batteries with a significantly higher energy density can also be used. For this reason, LiPo batteries are primarily used in aircraft models. Each cell provides $3.7\,\text{V}$. Thus a battery with two cells delivers

Figure 2.4: Battery pack with 2 × 4 NiMH cells and Tamiya plug

7.4 V, one with three cells 11.1 V. Special chargers, so-called *balancers*, which distribute the charging voltage to the individual cells are required for charging.

2.3 Mechanical Design

There are various drive concepts for mobile robots. The robot described here has two separate motors. This allows the robot not only to move forwards and backwards, but also to rotate on the spot. The platform to which the motors, sensors, etc. are attached is often circular (see Fig. 2.5). For our robot platform a diameter of approx. 15...20 cm (i.e., 6...8 in) is recommended.

In the simplest case, the platform can be made of wood. A robot platform made of acrylic is visually more appealing, but also more challenging in processing. The top side is typically reserved for the controller board and the LCD panel, but the battery pack can also be attached to the bottom side. The robot platform designed is shown in Fig. 2.6.

To give the robot platform stability, the two drive wheels must be supplemented by a third support point. The use of a transport or steering wheel as rear wheel is self-evident here (Fig. 2.7). Such casters are available in various sizes in DIY stores. In conjunction with the wheels from Solarbotics, a height of approx. 34 mm (i.e., 1.3 in) makes sense for the caster. However, the required

2.3 Mechanical Design

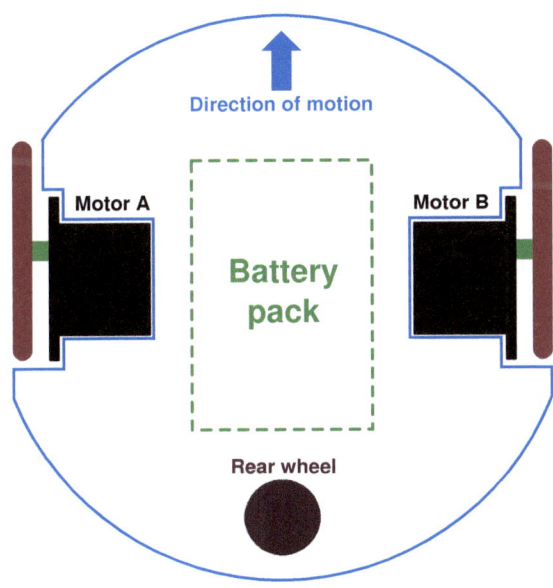

Figure 2.5: Robot platform

Figure 2.6: Robot platform with special servos and battery pack

overall height also depends on the mounting of the motors. On some mobile robots, e.g. the Pololu 3pi robot, a plastic ball is used for the third support point [1].

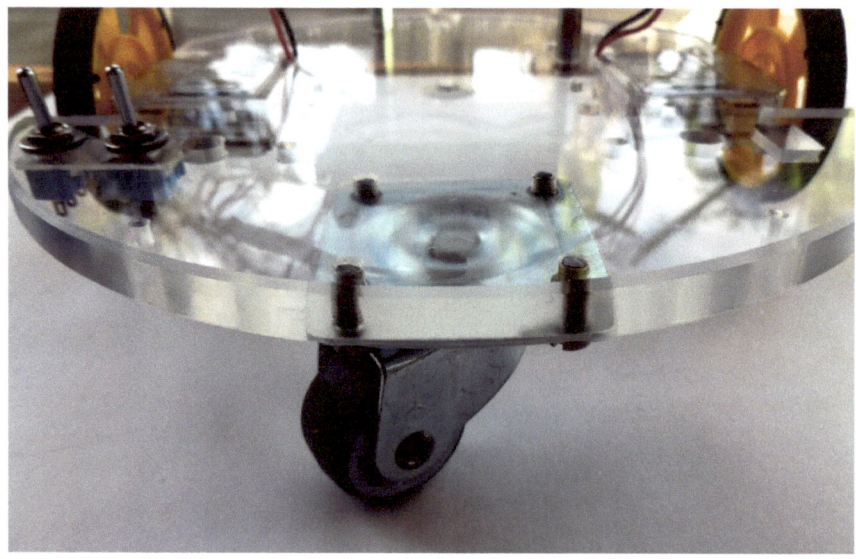

Figure 2.7: Rear wheel of the mobile robot

At least two switches should be provided as operating elements: One as the main switch, the other switch for the motors. The second switch can be used to test programs without the robot driving off and perhaps falling off the desk. In addition, some buttons could also be provided as control elements.

As an intermediate stage between a completely independent assembly and a complete robot kit one could also fall back on kits or ready-to-use devices for chassis or mobile robot platforms. Fig. 2.8 shows such a kit, which is sold under the designation ZK-2WD, where 2WD stands for two wheels drive. This kid contains all essential mechanical components including the gear motors. To use it with an Arduino board, however, the battery holder designed for 4 cells would have to be replaced by one for 6 cells (see Fig. 2.9).

2.3 Mechanical Design

Figure 2.8: Kit for a mobile robot chassis

Figure 2.9: Assembled robot chassis with battery holder for 6 cells

References

[1] *Pololu robotics and electronics.* `https://www.pololu.com/`.

[2] *Solarbotics.* `https://solarbotics.com/`.

[3] Wikipedia contributors: *Servo (radio control) — Wikipedia, the free encyclopedia.* `https://en.wikipedia.org/w/index.php?title=Servo_(radio_control)`.

Chapter 3

Motor Control

3.1 Speed Control with Pulse Width Modulation (PWM)

As a rule, the DC motors used in mobile robots should not always run at full nominal speed; instead, the speed of rotation should be continuously adjustable. There are basically two approaches. With analog control, a voltage drop is generated by a variable resistor (potentiometer) connected to the power supply voltage; with a digital control one would only use a switch (see Fig. 3.1). Both variants can also be built with transistors.

Figure 3.1: Control of DC motors: analog control with potentiometer (left), digital control with switch (right)

If the potentiometer is set such that only half the operating voltage is applied to the motor, the motor will operate approximately at half the speed. However, the potentiometer then consums the same electrical power as the motor. This reduces the efficiency. Furthermore, the energy impressed on the resistor causes the component to heat up (regardless of whether the resistor is designed as a potentiometer or a transistor).

The above-mentioned problems with an analog control can be avoided using a digital control, i.e., with switched operation. If the switch is open, no current flows. Theoretically, no voltage drops across a closed switch. In either case, there is no (noticeable) power consumption. Normally, a switch can only be used to select between stop and full speed. With a *pulse-width modulation (PWM)* you can also set (almost arbitrary) intermediate values [8]. For this purpose, the switch is quickly turned on and off with a frequency $f = 1/T$. In Fig. 3.2, t_{on} denotes the switch-on time and t_{off} the switch-off time, respectively, so that one oscillation period has the time $T = t_{\text{on}} + t_{\text{off}}$. The *duty ratio d* is the ratio of the switch-on time to the period duration and can assume values in the range from zero (always off) to one (always on). At a duty cycle of $d = 0.5$, the switch is closed for exactly the same time it is open. In this case half the voltage would be applied on average.

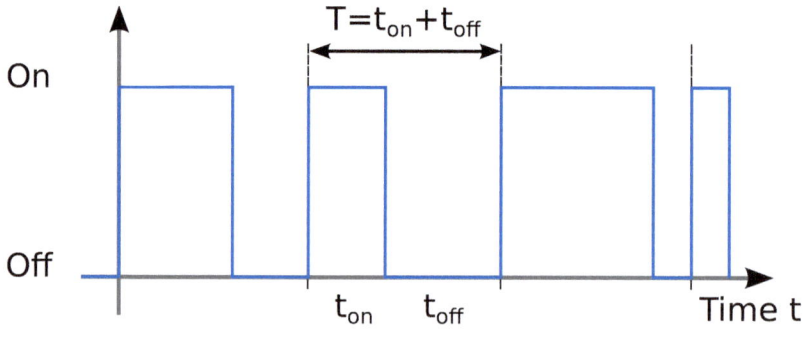

Figure 3.2: PWM signal

The control of the motors in the robot is slightly different, namely with a *bridge circuit*. It consists of four switches, which are realized as transistors (see Fig. 3.3). A bridge circuit allows the motor voltage to be reversed and thus enables the motor to operate in both directions as shown in Fig. 3.4.

Since the robot is driven by two DC motors, two bridge circuits are required accordingly. These are not assembled discretely (from individual components), but are realized with integrated circuits. Older motor shields often use the

3.1 Speed Control with Pulse Width Modulation (PWM)

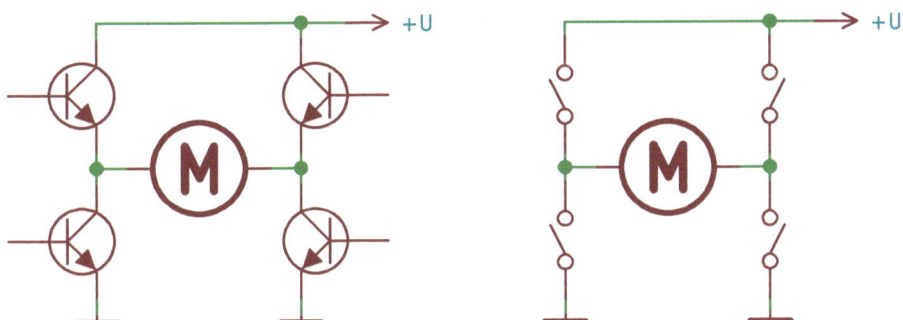

Figure 3.3: Bridge circuit: bridge with NPN transistors (left), bridge with switches (right)

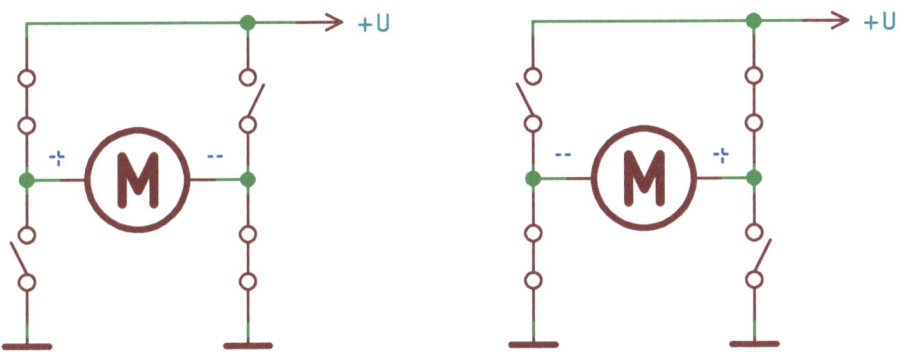

Figure 3.4: Change of direction with bridge circuit

integrated circuit L293, which is designed for currents up to 1 A [5]. Modern motor shields use the IC L298 provided for 2 A, which additionally allows the measurement of motor currents [4].

3.2 Arduino Motor Shield R3

The Arduino motor shield R3 (revision 3) shown in Fig. 3.5 is probably the most popular add-on module for motor control at the time of writing. The motor shield is intended for controlling two DC motors. Alternatively, it is also possible to control one stepper motor [1].

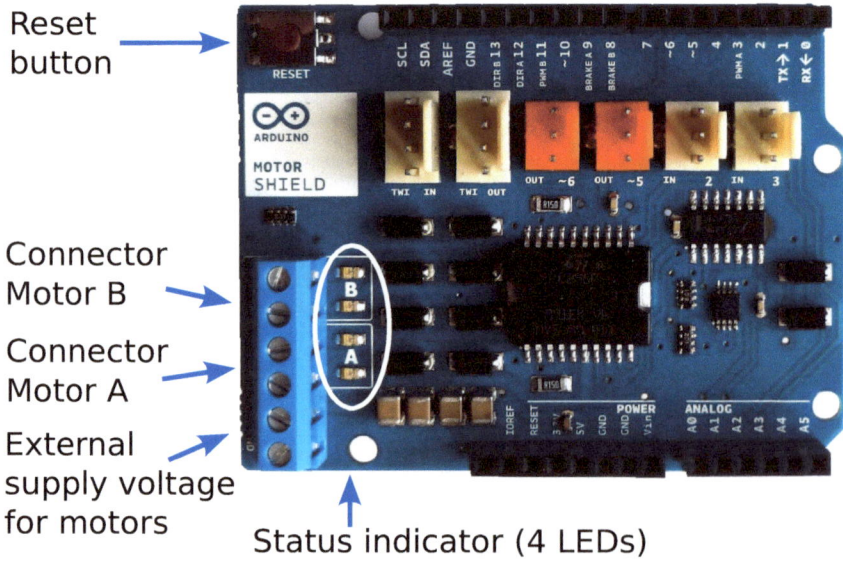

Figure 3.5: Arduino motor shield (component side)

Furthermore, it is also possible to brake both motors directly (instead of letting them run out) and measure the motor currents. If you do not need these additional functions, you can cut the corresponding solder bridges on the rear side and thus release the addresses (channels) provided for this purpose (see Fig. 3.6). With one further solder bridge it is possible to disconnect the power supply of the motors from the supply voltage V_{in} of the Arduino board in order to supply them with an external supply voltage V_{ext}.

3.2 Arduino Motor Shield R3

Figure 3.6: Arduino motor shield (conductor side)

The various functions of the motor shield can be addressed via several digital and analog channels. The corresponding (hard-wired) addresses are listed in Tab. 3.1.

Table 3.1: Addresses of the Arduino motor shield

Function or signal	Channel A	Channel B
Direction	12	13
Pulse width modulation (PWM)	3	11
Brake	9	8
Motor current measurement	A0	A1

The two channels A and B can each be used to control a DC motor. The direction of rotation is set via pins 12 and 13 of the Arduino board, the rotation speeds via the PWM outputs with pins 3 and 11, respectively. These addresses are hard-wired and can be defined as constants in the program as follows:

```
const byte DIRA = 12;
const byte DIRB = 13;
const byte PWMA = 3;
const byte PWMB = 11;
```

In the function **setup**, these four digital channels must be defined as outputs:

```
void setup()
{
  pinMode (DIRA, OUTPUT);
  pinMode (DIRB, OUTPUT);
  pinMode (PWMA, OUTPUT);
  pinMode (PWMB, OUTPUT);
}
```

In the main program, i.e., in the function **loop**, you would set or adjust the direction and speed of rotation of the motors according to your own requirements. In the next program example, the pins for the direction of rotation are set to LOW. The PWM output takes place via the function **analogWrite**, which expects values in the range from 0 to 255. The value 0 corresponds to the standstill of the motor, the value 255 provides the maximum speed. The value of 127 used here corresponds to approximately half of the maximum speed:

```
void loop() {
  // direction
  digitalWrite (DIRA, LOW);
  digitalWrite (DIRB, LOW);
  // half speed
  analogWrite (PWMA, 127);
  analogWrite (PWMB, 127); }
```

In our robot prototype, the left motor is controlled via channel A, the right motor via channel B. The motors are installed in the mobile robot with opposite orientation. When moving straight forward or backward, both wheels (right and left) should turn in the same direction, for which the motors must turn in opposite directions. The motors of the prototype shown in Fig. 3.7 were connected in such a way that both motors contribute to a forward movement if the direction of rotation is set according to the above-mentioned program.

3.2 Arduino Motor Shield R3

Figure 3.7: Robot prototype with Arduino motor shield

To stop the motors, either output the value zero as (quasi analog) PWM signal

```
analogWrite (PWMA, 0);
analogWrite (PWMB, 0);
```

or set the outputs (as digital signals) to LOW level:

```
digitalWrite (PWMA, LOW);
digitalWrite (PWMB, LOW);
```

To use the brakes, pins 8 and 9 must be declared as (digital) outputs. At LOW level the brake is inactive, at HIGH level it is actively braked. When using gear motors, an electronic brake is hardly necessary; the normal run-out of the motor is barely noticeable due to the reduction of the gear. If you do not use the brake, you do not have to configure pins 8 and 9: Two pull-down resistors on the motor shield pull the corresponding pins to ground and thus to LOW level. If you want to use pins 8 or 9 for other purposes, you have to cut the corresponding solder bridges on the back side of the motor shield (see Fig. 3.6).

3.3 Velleman Motor Shield

Another popular motor shield is offered by Velleman (see Fig. 3.8). This Motor & Power Shield is available both as a kit (KA03) and as a fully assembled board (VMA03) [6, 7]. The shield also allows the control of two DC motors or one stepper motor.

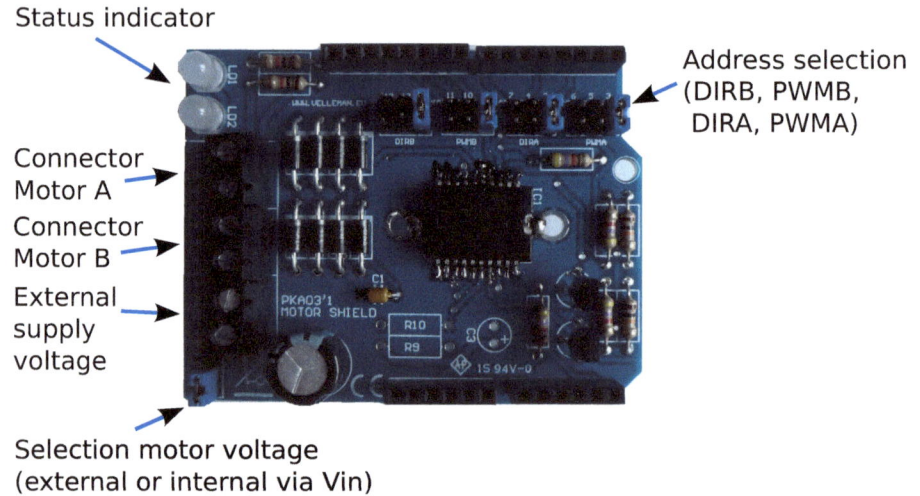

Figure 3.8: Motor shield KA03 from Velleman

Various digital channels of the Arduino board are available for selecting the directions (DIRA, DIRB) or the PWM control (PWMA, PWMB). The channels are selected via jumpers (see Tab. 3.2). A further jumper allows the selection of the motor voltage, which is provided either as external voltage V_{ext} or internally via the supply voltage V_{in} of the Arduino board. The utilization of an external voltage V_{ext} is useful for motors operating at 24 V, for example. Regardless of the voltage source selected, the motor voltage can be measured via analog channel A5 (see Section 6.5 on p. 73). For this purpose, a voltage divider consisting of resistors R9 and R10 is provided on the PCB, whereby the measured signal is additionally smoothed with the capacitor C3. These components were not fitted to the assembled board shown in Fig. 3.8.

Three possibilities are provided per channel for direction of rotation and PWM control. This could be used to connect up to three motor shields on top

3.3 Velleman Motor Shield

Table 3.2: Addresses of the Velleman motor shield

Signals	Channels
DIRA	2, 4, 7
DIRB	8, 12, 13
PWMA	3, 5, 6
PWMB	9, 10, 11
Motor voltage	A5

of each other and thus control up to 6 DC motors. To perform a first test, we want to be content with a (single) motor shield and make the pin or address selection shown in Fig. 3.9.

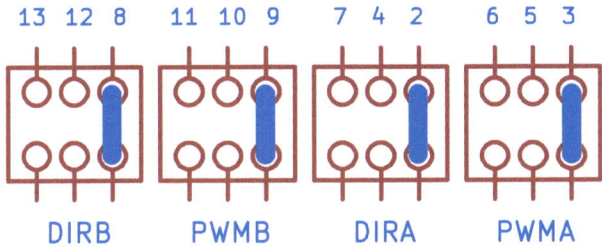

Figure 3.9: Standard address selection for the Velleman motor shield

This pin assignment would be embedded in the program like that:

```
const byte DIRA = 2;
const byte DIRB = 8;
const byte PWMA = 3;
const byte PWMB = 9;
```

With this address assignment you can use the program created for the Arduino motor shield for testing the Velleman motor shield. To show the direction of rotation, the two-color LEDs LD1 and LD2, which light green or red, respectively, are provided on the Velleman motor shield.

The different pin assignments between the official Arduino motor shield and the motor shield from Velleman not only affect the actual motor control, but also the channels intended for other purposes (e.g. for the LCD module, buttons, sensors, etc.). Since the pins of the official Arduino motor shield R3 are hardwired, it makes sense to adapt the address selection of the Velleman motor shield

Figure 3.10: Robot prototype with Velleman Motor Shield

to the assignment of the Arduino motor shield. For three of the four signals, the corresponding pin selection is immediately possible with the jumpers provided (Fig. 3.11). Only for setting the direction of rotation of channel A with the signal DIRA a separate connection must be provided.

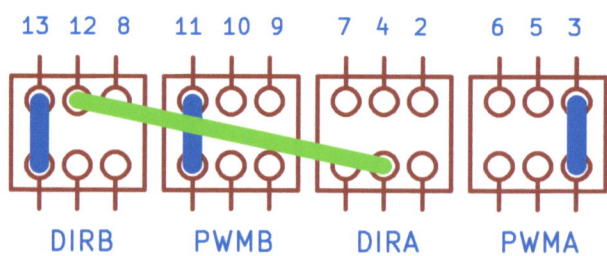

Figure 3.11: Address selection for the Velleman motor shield in accordance with the Arduino motor shield

3.4 C++ Class for Motor Control

If you want to control both motors for the movement of the mobile robot, four function calls are necessary when using elementary Arduino commands (two of them with `digitalWrite` for direction setting and two of `analogWrite` for PWM output). In this section, a C++ class is created step by step for easy handling of the motor control. In Section 3.5 the resulting class is integrated into the Arduino system as a library.

First, we declare a minimum version of the `Motor` class:

```cpp
class Motor {
// pins for direction
 byte DIRA, DIRB;
// pins for PWM
 byte PWMA, PWMB;
 public:
 Motor (byte, byte, byte, byte);
};
```

The class contains the attributes DIRA, DIRB, PWMA and PWMB, which characterize the pins required for controlling the motor shield. The specific addresses are assigned with the constructor, the call of which requires four arguments of type `byte`. The values passed during the call are assigned to the above-mentioned attributes:

```cpp
Motor::Motor (byte pinDIRA, byte pinDIRB,
              byte pinPWMA, byte pinPWMB)
{
  DIRA = pinDIRA;
  DIRB = pinDIRB;
  PWMA = pinPWMA;
  PWMB = pinPWMB;
}
```

Alternatively, you can implement these assignments via an initialization list (see e.g. [3]). In addition, both motors are switched off as well:

```cpp
Motor::Motor (byte pinDIRA, byte pinDIRB,
              byte pinPWMA, byte pinPWMB)
: DIRA(pinDIRA), DIRB(pinDIRB),
```

```
  PWMA(pinPWMA), PWMB(pinPWMB)
{
  digitalWrite (PWMA, 0);
  digitalWrite (PWMB, 0);
}
```

In your own program code you would create an instance (here: `motor`) of the class `Motor` and pass the pins relevant for motor control. The values given in this example correspond to the address assignment of the official Arduino motor shield R3:

```
Motor motor(12, 13, 3, 11);
```

This very common pin assignment could also be provided for the standard constructor to be called without arguments, for example. You could specify default values for the above constructor in the class definition:

```
Motor (byte=12, byte=13, byte=3, byte=11);
```

The instantiation without arguments would create an object of the class with the stored standard assignments. The existence of a standard constructor is also useful if you want to create a class derived from the class `Motor`:

```
Motor motor;
```

The required pins are now stored in the class, but still have to be configured as outputs. To do this, you extend the class `Motor` by the method `begin`, which has neither an argument nor a return value:

```
void Motor::begin()
{
  pinMode (DIRA, OUTPUT);
  pinMode (DIRB, OUTPUT);
  pinMode (PWMA, OUTPUT);
  pinMode (PWMB, OUTPUT);
}
```

The method `begin` must be called in the function `setup`:

```
void setup()
{
  motor.begin();
}
```

Next, we create the method `setValues` to set direction and speed. In the class declaration, this method is marked with `private` so that it can only be used within the class.

```
private:
void setValues (byte, byte, int);
```

The class declaration is additionally extended by the two attributes `Vmax` and `Vmin`, which describe the maximum and minimum values for the PWM and thus for angular velocity and speed. With an 8-bit PWM only values from 0 to $2^8 - 1 = 255$ are possible. If you do not wish to operate the motors (permanently) with the maximum operating voltage, you can specify a smaller value for `Vmax`, e.g. 127 for an average half operating voltage. Conversely, if the voltage or duty cycle is too low, the motor typically no longer rotates; the current is only converted into heat. This case is also undesirable.

```
// PWM values for max and min speed
static const byte Vmax = 255;
static const byte Vmin = 40;
```

Three values must be passed to the method `setValues`. In addition to the relevant pins, the direction and speed of rotation is set via the variable `speed`. The sign of the argument `speed` indicates the direction of rotation. A positive value for the forward rotation and a negative value for the backward rotation are provided. The absolute value of the variable `speed` corresponds to the speed set via the PWM, which is limited upwards by `Vmax` and is set to zero for values below `Vmin`.

```
void Motor::setValues (byte pinDir, byte pinPwm, int speed)
{
  // direction
  digitalWrite (pinDir, speed>=0 ? LOW : HIGH );
  // speed
  speed=abs(speed);
  speed=min(speed,Vmax);
  if (speed<Vmin) speed=0;
  analogWrite(pinPwm, speed);
}
```

With the method `setValues` you are able to control the two motors taking into account the pin assignment stored as class attributes. The methods `writeA` and `writeB` are defined directly in the class declaration as inline functions:

```cpp
// motor A: left
void writeA (int speed)
   { setValues (DIRA, PWMA, speed); };
// motor B: right
void writeB (int speed)
   { setValues (DIRB, PWMB, speed); };
```

The method `write` can be used to influence both motors with a single function call:

```cpp
void Motor::write (int speedA, int speedB)
{
  writeA(speedA);
  writeB(speedB);
}
```

Obviously, the corresponding prototype of this function must also be listed in the class declaration. Make sure that the functions `writeA`, `writeB` and `write` are declared as public (in terms of their access rights). With the following test program, the mobile robot shows the different maneuvers:

```cpp
int V=200; // PWM value
int T=500; // time in ms
void loop()
{
  // forward
  motor.write ( V, V); delay(T);
  // right
  motor.write ( V,-V); delay(T);
  // backward
  motor.write (-V,-V); delay(T);
  // left
  motor.write (-V, V); delay(T);
}
```

3.5 Motor Control Library

The motors of a mobile robot will probably be used in almost every application. In order not to have to copy the corresponding code for the motor control every time, it makes sense to create a library of the `Motor` class provided in the previous section.

With the installation of the Arduino IDE, a folder for own program sketches – `sketchbook` – is created. In addition to the programs you wrote yourself, the folder also contains a directory called `libraries`, which initially contains no further entries. In the directory `libraries` we create the new folder `Motor`, which should contain the files for the motor library (see [2]).

The file `Motor.h` is a so-called *header file* which contains the type definitions of the classes or functions contained in the library. The sequence

```
#ifndef Motor_h
#define Motor_h
...
#endif
```

prevents this definition file from being processed several times (by mistake). Declarations specific to the Arduino environment (such as the `digitalWrite` function) are included via the `Arduino.h` header file.

```
#include "Arduino.h"
```

Together with the declaration of the class `Motor` you obtain the following header file, which is to be saved in the above mentioned folder `Motor`:

```
/*
  Motor.h - Library for motor control
  Created by Klaus Röbenack, 2015
*/

#ifndef Motor_h
#define Motor_h

#include "Arduino.h"

class Motor {
  protected:
```

```cpp
  // pins for direction
  byte DIRA, DIRB;
  // pins for PWM
  byte PWMA, PWMB;
  // max/min values for PWM
  static const byte Vmax = 255;
  static const byte Vmin = 40;
  public:
  // constructor
  Motor (byte=12, byte=13, byte=3, byte=11);
  // initialization
  void begin();
  // Motor A: left
  void writeA (int speed) { setValues (DIRA, PWMA, speed); };
  // Motor B: right
  void writeB (int speed) { setValues (DIRB, PWMB, speed); };
  // Motor A and B
  void write (int, int);
  private:
  void setValues (byte, byte, int);
};

#endif
```

Contrary to the class declaration given in the previous section, the keyword **protected** has now been inserted at the beginning. Direct access to the subsequent class elements is then no longer possible from outside. However, derived classes can access these elements.

While the header file `Motor.h` only contains the declarations, the actual implementation takes place via the C++ file `Motor.cpp`, which also has to be stored in the `Motor` folder:

```cpp
/*
  Motor.cpp - Library for motor control
  Created by Klaus Röbenack, 2015
*/

#include "Arduino.h"
#include "Motor.h"
```

3.5 Motor Control Library

```cpp
// constructor
Motor::Motor (byte pinDIRA, byte pinDIRB,
              byte pinPWMA, byte pinPWMB)
: DIRA(pinDIRA), DIRB(pinDIRB), PWMA(pinPWMA), PWMB(pinPWMB)
{
  digitalWrite (PWMA, 0);
  digitalWrite (PWMB, 0);
}

// initialization
void Motor::begin()
{
  pinMode (DIRA, OUTPUT);
  pinMode (DIRB, OUTPUT);
  pinMode (PWMA, OUTPUT);
  pinMode (PWMB, OUTPUT);
}

// internal function
void Motor::setValues (byte pinDir, byte pinPwm, int speed)
{
  // direction
  digitalWrite (pinDir, speed>=0 ? LOW : HIGH );
  // speed (PWM value)
  speed=abs(speed);
  speed=min(speed,Vmax);
  if (speed<Vmin) speed=0;
  analogWrite(pinPwm, speed);
}

// speed for both motors
void Motor::write (int speedA, int speedB)
{
  writeA(speedA);
  writeB(speedB);
}
```

To test the library, we create a small, independent program from the remaining code snippets of the previous section. After restarting the Arduino environment, the library should be listed under **Sketch > Include Library**. If you select the library `Motor`, the corresponding header file (here: `Motor.h`) will be included in the source code:

```
// test of the motor library
#include <Motor.h>

// create an instance of the class Motor
Motor motor(12, 13, 3, 11);

void setup()
{
  motor.begin();
}

int V=200; // PWM value
int T=500; // time in ms

void loop()
{
  // forward
  motor.write ( V, V); delay(T);
  // right
  motor.write ( V,-V); delay(T);
  // backward
  motor.write (-V,-V); delay(T);
  // left
  motor.write (-V, V); delay(T);
}
```

This example code can also be stored directly with the library. To do this, create a directory named `Examples` in the `Motor` folder and store the example code in a new subdirectory, e.g. `Motor_Test`. More than one example can be provided for a library. After the next start of the Arduino environment, the example can be found in the general collection of examples, namely under **File > Examples > Motor > Examples > Motor_Test**.

There are also various possibilities to extend the library. For example, special functions or methods could be provided for the usual basic maneuvers (forward and reverse, clockwise and anti-clockwise rotation). Instead of directly extending the motor library, the additional functions could also be implemented in a derived class (see also Sections 7.3 to 7.5 and Section 8.2).

References

[1] *Arduino Motor Shield.* https://arduino.cc/en/Main/ArduinoMotorShieldR3.

[2] *Writing a Library for Arduino.* https://arduino.cc/en/Hacking/LibraryTutorial.

[3] Josuttis, N. M.: *Object-Oriented Programming in C++.* John Wiley & Sons, 2002.

[4] STMicroelectronics: *L298; Dual Full-Bridge Driver*, 2000. Datasheet.

[5] Texas Instruments: *L293, L293D; Quadruple Half-H Drivers.* Datasheet.

[6] Velleman: *KA03 Motor & Power shield Arduino®.* Illustrated assembly manual.

[7] Velleman: *VMA03 Motor & Power shield Arduino®.* Manual.

[8] Wikipedia contributors: *Pulse-width modulation — Wikipedia, the free encyclopedia.* https://en.wikipedia.org/w/index.php?title=Pulse-width_modulation.

Chapter 4

LCD Output

4.1 Connecting an LCD Module

For the control of alphanumeric LCD modules, the controller HD44780 developed by the Hitachi has de facto become an industry standard [1]. This integrated circuit and its largely or fully compatible clones and redesigns are nowadays used to control LCD modules of various sizes (see Fig. 4.1). Common display formats are 8×2, 16×2 to 20×4 (columns × rows).

Figure 4.1: LCD module with 16 columns and 2 rows (16×2)

For the prototype robot (comparatively large) display modules with 20 columns and 4 lines were used. Most display moduls also have the same pin assignment as shown in Tab. 4.1.

Table 4.1: Typical pin and address assignment of LCD modules

Pin number	Symbol	Function or signal
1	V_{SS}	Ground (0 V)
2	V_{DD}	Supply voltage (+5 V)
3	V_0	Contrast adjustment
4	RS	Register selection (command: 0, data: 1)
5	R/W	Read/Write (write: 0, read: 1)
6	E	Enable
7-14	DB0-DB7	8-bit Data bus
15	A	Backlight supply voltage
16	K	Backlight ground

To control the display module, 8 data bits are provided in conjunction with some control signals. In addition to 8-bit operation, 4-bit operation is also possible, which requires fewer pins. Typically, the R/W pin is not used, so that only 6 digital channels are necessary. If one considers the digital channels already occupied by the Arduino motor shield, then the assignment scheme specified in Tab. 4.2 would be conceivable.

Table 4.2: Assignment of the digital channels of the Arduino board for interfacing an LCD module

Digital channels (Arduino)	Signal on the LCD module
10	RS
2	E (Enable)
7	D4
6	D5
5	D6
4	D7

When connecting an LCD module, ground (GND) and supply voltage (5 V) must be taken into account in addition to the data lines just mentioned (see Fig. 4.2). The contrast of the display can be adjusted using a $10\,\mathrm{k\Omega}$ trimmer potentiometer. If necessary, the backlight must be taken into account. Typically this is realized with LEDs and requires a series resistor of a few ohms [3].

4.1 Connecting an LCD Module

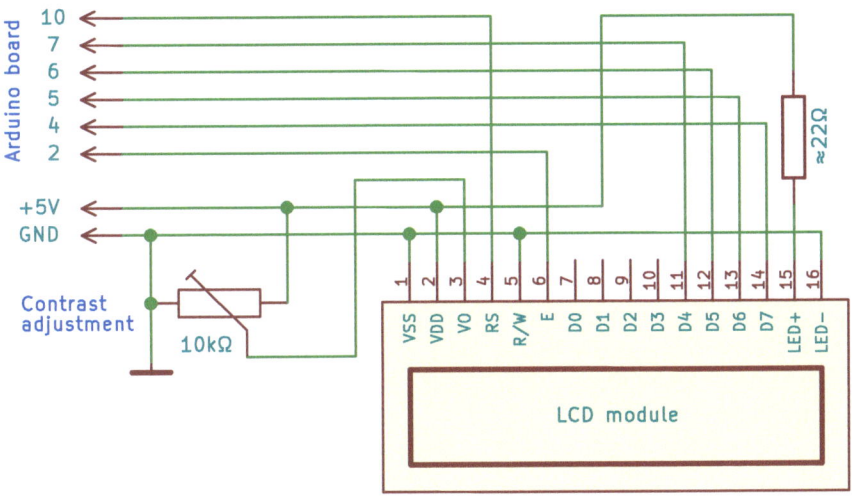

Figure 4.2: Wiring of the LCD module

Fig. 4.3 shows a prototype of the robot with the 20×4 display AV2040 from ANAG VISION [3].

Figure 4.3: Robot with mounted LCD module

4.2 Control of the LCD Module

To control the LCD module, we have to include the library for LCDs, which provides the class `LiquidCrystal` [4].

```
#include <LiquidCrystal.h>
```

Then we create the object `lcd` as an instance of the class `LiquidCrystal`. The pins used for wiring with the Arduino board also need to be specified. The order of the signals used must be observed (RS, Enable, D4, D5, D6, D7):

```
LiquidCrystal lcd(10, 2, 7, 6, 5, 4);
```

The LCD module must be initialized with the function `setup`. The class `LiquidCrystal` provides the method `begin`, where the number of rows and columns of the module must be passed, e.g. 20 columns and 4 rows:

```
lcd.begin(20, 4);
```

The method `print` is used for the output. Admissible data types are `char`, `byte`, `int`, `long` or `string`.

```
lcd.print("LCD test!");
```

The entire program for testing the LCD module will then look like this:

```
// LCD library
#include <LiquidCrystal.h>

// create object
LiquidCrystal lcd(10, 2, 7, 6, 5, 4);

void setup() {
  // column and row number
  lcd.begin(20, 4);
  // text output
  lcd.print("LCD test!");
}

void loop() {}
```

The `LiquidCrystal` class contains numerous other methods for controlling and formatting the output. The most important functions (from the author's point of view) are listed in Tab. 4.3.

Table 4.3: Important methods of the class `LiquidCrystal` [4]

Method, calling convention	Description
`begin(x,y)`	Initialize the display with x columns and y rows
`clear()`	Clears the screen and sets the cursor in the upper left corner
`home()`	Sets the cursor in the upper left corner
`setCursor(x,y)`	Sets cursor to column x and line y (starting from zero)
`write(c)`	Output of a single character (`char`)
`print(data)`	Output for various data types
`print(data,base)`	Prints a number w.r.t. a specified base (predefined constants: BIN, OCT, DEC, HEX)

4.3 Formatting the Output

To format the output you can use the standard C/C++ library `stdio.h` [2]. To use this library you have to include the corresponding header file:

```
#include <stdio.h>
```

The output on a PC would be done with the `printf` function, which is not directly available for output on an LCD display. Alternatively, you can use the very similar function `sprintf`. To do this, declare a character string of the desired length as an array of type `char`. The end of a character string is marked internally with a zero. For the LCD module with 20 columns per line, an array of length 21 is needed:

```
char line[21];
```

The function `sprintf` does not make the output itself, but stores the result in the string passed as pointer in the first argument (here: `line`). The second argument when calling `sprintf` is a special string that contains not only "normal" characters but also format specifiers for the output of different data types. The respective output is introduced with the character "%", followed, if necessary, by a number indicating the length. The other function arguments are the values to be output.

```
sprintf(line,"Integer: %4d",value);
lcd.print(line);
```

The method **print** is used for the output on the LCD. Unfortunately, the output or formatting of floating point numbers did not work with the function **sprintf** on the Arduino platform. Tab. 4.4 contains the common placeholders in the format string.

Table 4.4: Important placeholders in the format string of the function **sprintf**

Placeholder	Description
%d, %i	Integer numbers, possibly with sign
%u	Nonnegative integer numbers
%x, %X	Hexadecimal output of nonnegative integer numbers
%c	Character (**char**)
%s	String (null-terminated array of **char**)

However, you can use the **sprint** function to output floating point numbers indirectly. This is done by converting the floating point number with the function **dtostrf** of the AVR standard library [5] into a character string that can then be processed further with the formatting specification "**%s**" from **sprintf**. In addition to the floating point number, the number of characters in the output (including the decimal point), the number of digits after the decimal point, and a field of type char for the result must be passed to function **dtostrf**:

```
float f=3.14;     // floating point number
char str_num[5];  // array for converted floating point number
dtostrf(f, 4, 2, str_num);
sprintf(line, "Number: %s",str_num);
```

The method **print** of the class **LiquidCrystal** allows the output of different data types, but must be called separately (and thus multiple times) for each type. For example, a combined output of text with numbers of the types **int** and **float** could look like this:

4.3 Formatting the Output

```
int i=4;
float f=3.14;
lcd.print("int: ");
lcd.print(i);
lcd.print(" float: ");
lcd.print(f);
```

In C++ the data streams `cin` and `cout` are available for input and output via the console, which enable the output of different data types in a very simple manner. A similar functionality is provided by the library **Streaming** for the Arduino platform [6].

The library **Streaming** is not included in the libraries provided with the Arduino IDE. However, the Arduino website explicitly refers to this library. After downloading the corresponding ZIP file, unpack it in the usual way in the subdirectory `libraries` of the folder `sketchbook`. By calling the Arduino IDE again, the library should be recognized automatically. Use Sketch > Import Library to integrate the library into your own program. The header file is inserted into the source code:

```
#include <Streaming.h>
```

For the above created object `lcd` you can now send different data types to the outputs with the insertion operator "<<", e.g.

```
lcd<<"int: "<<i<<" float: "<<f;
```

The library **Streaming** translates this line into individual `print` statements and thus does not occupy any separate memory at runtime.

References

[1] Wikipedia contributors: *Hitachi HD44780 LCD controller — Wikipedia, the free encyclopedia.* `https://en.wikipedia.org/w/index.php?title=Hitachi_HD44780_LCD_controller`.

[2] Wikipedia contributors: *Printf format string — Wikipedia, the free encyclopedia.* `https://en.wikipedia.org/w/index.php?title=Printf_format_string`.

[3] Anag Vision: *AV2040.* Datasheet.

[4] *Arduino — LiquidCrystal.* `https://www.arduino.cc/en/Reference/LiquidCrystal`.

[5] *avr-libc: <stdlib.h>: General utilities.* `https://www.nongnu.org/avr-libc/user-manual/group__avr__stdlib.html`.

[6] *Streaming | Arduiniana.* `http://arduiniana.org/libraries/streaming/`.

Chapter 5

Buttons and Switches

5.1 Connecting Push-Buttons

Buttons or switches attached to the robot can be used to select the respective program mode. In this chapter, buttons are used to detect limitations of the robot's range of motion. To read the states of the buttons, the digital channels are used, although many of them are already occupied by the motor shield or the LCD display. Available addresses would be the digital channels 0 and 1. These pins are also needed for program transmission and/or serial communication on the Arduino Uno board, but this does not prevent the intended use.

```
const byte pinButtonA = 0;
const byte pinButtonB = 1;
```

If you don't use the brake function on the Arduino motor shield, you could also use pins 8 and 9. To do this you would have to cut the solder bridges marked BRAKE DISABLE on the back side of the motor shield (see Fig. 3.6 on p. 21).

```
const byte pinButtonA = 8;
const byte pinButtonB = 9;
```

To read the state of the buttons or switches with a digital input, a defined signal level must be guaranteed for each position of the switch. This can be done, for example, via pull-up or pull-down resistors as shown in Fig. 5.1. In the case of a pull-up resistor, the operating voltage and thus the HIGH level is present when the switch is open; a closed switch delivers the LOW level. With a pull-down resistor, the conditions are reversed: when the switch is open, the resistor pulls the input to LOW level; when the switch is closed, the operating voltage delivers HIGH level.

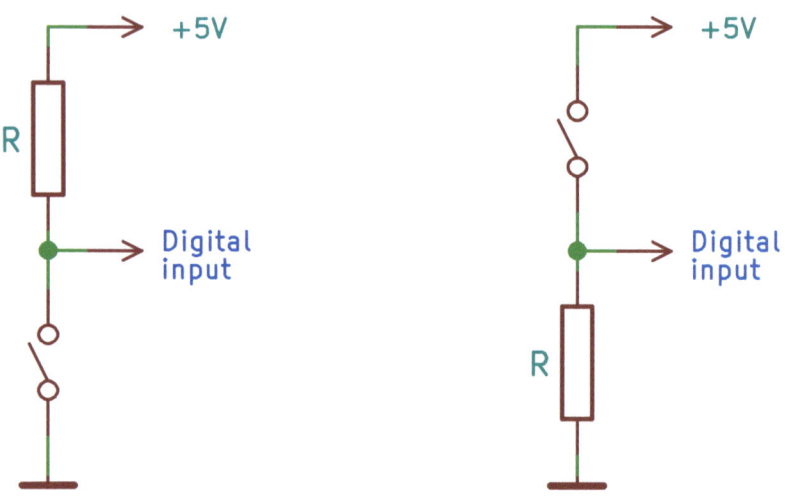

Figure 5.1: Push-button with pull-up resistor (left) or pull-down resistor (right)

In the example provided with the Arduino development environment, which can be called via File > Examples > 02.Digital > Button, a button to 5 V and a 10 kΩ pull-down resistor to ground are wired to pin 2. The retrieved signal is sent to the LED at pin 13 in the example program.

We want to omit additional resistors. The Arduino boards are equipped with pull-up resistors in the microcontroller which can be switched on if required. In the function `setup`, we assign the input mode with pull-up resistors to the above mentioned pins [2]:

```
pinMode (pinButtonA, INPUT_PULLUP);
pinMode (pinButtonB, INPUT_PULLUP);
```

The buttons must then be connected to ground from pins 0 and 1 (or, alternatively, from pins 8 and 9) as shown in Fig. 5.2.

5.1 Connecting Push-Buttons

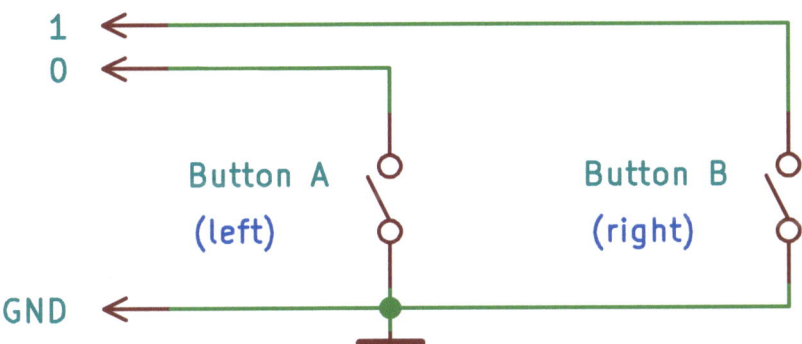

Figure 5.2: Two buttons connected to the Arduino board

The states of the digital channels 0 and 1 are read with the function `digitalRead`, to which the number of the corresponding pin is to be transferred. At LOW level this function returns 0, at HIGH level 1. These values are declared as constants LOW and HIGH in the Arduino file `Arduino.h`, see [3]. If the LCD module has been initialized in the usual way, the result of the buttons input pins can be displayed in the program as follows:

```
void loop()
{
  lcd.setCursor(0,0);
  lcd.print("Button A: ");
  lcd.print(digitalRead(pinButtonA));
  lcd.setCursor(0,1);
  lcd.print("Button B: ");
  lcd.print(digitalRead(pinButtonB));
}
```

Microswitches with long lever and rollers at the end of the lever are recommended for the intended use of obstacle detection. With the prototype robot, the mounting was done on a small acrylic plate as shown in Fig. 5.3. The acrylic plate with the microswitches is mounted on the front of the robot (see Fig. 5.4).

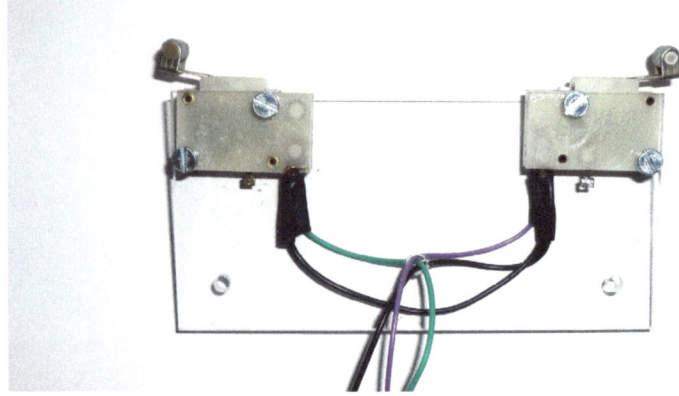

Figure 5.3: Mounting of the two microswitches on an acrylic plate

Figure 5.4: Mobile robot with microswitches on the front side

When reading the state of a push-button with a pull-up resistor, the return value 0 obtained from the function `digitalRead` corresponds to a pressed push-button, the value 1 to an open push-button. With two buttons, four states are possible (none pressed, button A or button B pressed, both pressed). You could query these states with nested `if` statements or `if-elseif` combinations. From the author's point of view it is more elegant to combine the two query results in a new integer variable (here: **buttons**):

```
buttons  =    digitalRead(pinButtonA);
buttons += 2*digitalRead(pinButtonB);
```

The return value of push-button A is added to the return value of push-button B scaled by a factor of 2. The four button states are mapped to the numbers 0 to 3. The corresponding reaction of the robot to the respective state of the two buttons can then be implemented very easily using the instruction `switch`. This approach is used in the following main loop to display the respective button assignment on the LCD panel:

```
void loop()
{
  lcd.home();
  lcd.print("Buttons: ");
  buttons  =    digitalRead(pinButtonA);
  buttons += 2*digitalRead(pinButtonB);
  switch(buttons) {
    case 0: lcd.print("Both "); break;
    case 1: lcd.print("Right"); break;
    case 2: lcd.print("Left "); break;
    case 3: lcd.print("None "); break;
  }
}
```

5.2 Obstacle Avoidance using Push-Buttons

In this section, we want to create a program that detects obstacles based on the states of the buttons and performs suitable evasive maneuvers. For this purpose, the motor is to be controlled and the respective maneuvers are to be displayed on the LCD module. Therefore both libraries — `LiquidCrystal` and `Motor` — have to be included:

```
// libraries: LCD and motor
#include <LiquidCrystal.h>
#include <Motor.h>
```

Then, you must create an instance of each of the corresponding classes:

```
// create objects
LiquidCrystal lcd(10, 2, 7, 6, 5, 4);
Motor motor(12, 13, 3, 11);
```

The next step is to define the pins for the buttons and create an integer variable for the combined state:

```
const byte pinButtonA = 0;
const byte pinButtonB = 1;
int buttons;
```

The function `setup` initializes the LCD module and motor control. In addition, the input channels used to read the buttons are configured with the internal pull-up resistors.

```
void setup() {
  lcd.begin(20, 4);
  motor.begin();
  // button input pins
  pinMode (pinButtonA, INPUT_PULLUP);
  pinMode (pinButtonB, INPUT_PULLUP);
}
```

After this preparation we can now turn our attention to the real problem, namely a programme to avoid obstacles. Normally, the robot moves straight ahead at the speed defined by the value `Speed`. When the robot comes into contact with an obstacle, it should first move back a bit. If the left button (button A) registers an obstacle, then the robot should turn approx. 90° to the right, in the case of the right button (button B) to the left. If both buttons are pressed simultaneously, i.e., if the robot hits an obstacle at the front, a turning maneuver (180° rotation) must be performed. The time intervals intended for reversing, lateral rotation or turning maneuvers must be determined

5.2 Obstacle Avoidance using Push-Buttons

experimentally and specified in ms under `T_back`, `T_rot` or `T_turn`.

```
// parameters for the maneuvers
int Speed = 180;
int T_back =   800;
int T_rot  =   500;  //  90 deg rotation
int T_turn = 1000;  // 180 deg rotation
```

The described evasion strategy is implemented in the main loop of the program:

```
void loop() {
  lcd.home();
  buttons  =   digitalRead(pinButtonA);
  buttons += 2*digitalRead(pinButtonB);
  if (buttons==3) {
    // move straight ahead
    lcd.print("Forward    ");
    motor.write (Speed, Speed);
  }
  else {
    // move back
    lcd.print("Backward   ");
    motor.write (-Speed,-Speed);
    delay(T_back);
    lcd.home();
    // evasive maneuvers
    switch(buttons) {
      case 0: lcd.print("Turn       ");
        motor.write (Speed,-Speed); delay(T_turn); break;
      case 1: lcd.print("To the left ");
        motor.write (-Speed,Speed); delay(T_rot); break;
      case 2: lcd.print("To the right");
        motor.write (Speed,-Speed); delay(T_rot); break;
    }
  }
}
```

5.3 Wired Remote Control with Switches

This section describes a simple wired remote control of the mobile robot. This requires two change-over switches with a stable off position in the centre, whereby no outward connection is required for the central position. On the console used for remote control, the switches may be arranged as shown in Fig. 5.5.

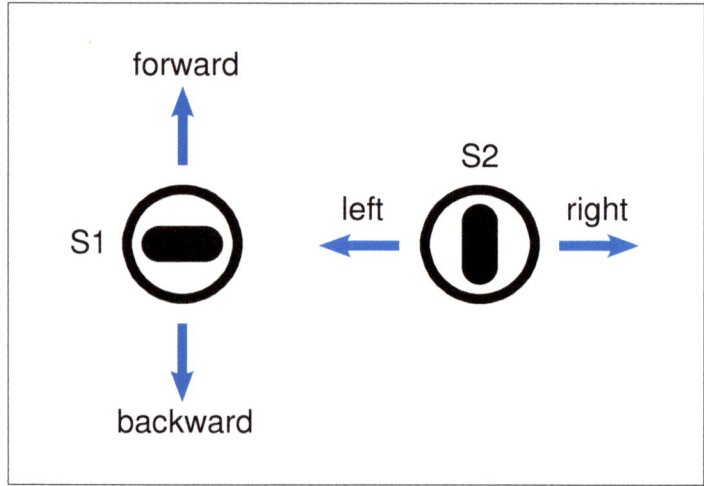

Figure 5.5: Arrangement of switches for wired remote control

Four digital inputs are required to query the change-over switches. In the following configuration the digital channels 0, 1, 8 and 9 are used. The schematic is shown in Fig. 5.6.

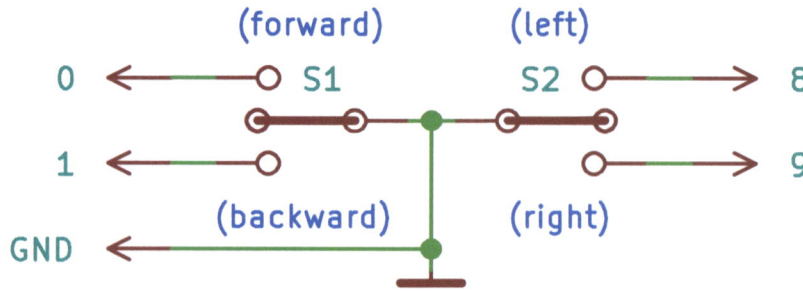

Figure 5.6: Wiring of change-over switches for remote control

5.3 Wired Remote Control with Switches

Please note that digital channels 8 and 9 can only be used for remote control if they are not already occupied by the brake function of the Arduino motor shield. With the Arduino Leonardo board you could alternatively use some of the connections of the ICSP adapter as additional digital inputs (see Section 5.4). The remote control must be connected to the robot (ground and four signal lines) via a five-core flexible cable.

In the software, the numbers of the used digital channels are stored in a one-dimensional array:

```
const byte pinButton[] = {0, 1, 8, 9};
```

These channels are initialized as digital inputs with pull-up resistor in the function `setup`. The dimension of the above mentioned array is determined by the `sizeof` operator. We assume that the LCD panel is integrated in the usual way (calling header file, creating the object `lcd` of the class `LiquidCrystal`).

```
void setup() {
  lcd.begin(20, 4);
  // button input pins
  for (byte i=0; i<sizeof(pinButton); i++)
    pinMode(pinButton[i],INPUT_PULLUP);
}
```

The configuration with internal pull-up resistors gives the value 1 when reading the digital inputs with the switch open and 0 when the switch is closed; the correct wiring of the four switches can be checked on a four-line LCD panel with a simple main loop:

```
void loop() {
  lcd.clear();
  for (byte i=0; i<=3; i++) {
    lcd.setCursor(0,i);
    lcd.print("Pin ");
    lcd.print(pinButton[i]);
    lcd.print(": ");
    lcd.print(digitalRead(pinButton[i]));
  }
  delay(100);
}
```

For further processing, we combine the four bits of the switches into an integer value using the function `ReadButtonState`:

```
int ReadButtonState () {
  int val = 0;
  for (byte i=0; i<=3; i++)
    val = (val<<1) + digitalRead(pinButton[i]);
  return val;
}
```

With the following main loop, the push-button values are cyclically displayed:

```
void loop() {
  lcd.clear();
  lcd.print("State: ");
  lcd.print(ReadButtonState());
  delay(100);
}
```

If both change-over switches are in the middle position, all four bits are set. In this case, the robot should be at a standstill. Exactly one bit is reset in each of the four main directions of movement. The relevant combinations are shown in Tab. 5.1.

Table 5.1: States of the change-over switches

Value	Bit 3	Bit 2	Bit 1	Bit 0	Maneuver
15	1	1	1	1	stop
Main directions					
7	0	1	1	1	forward
11	1	0	1	1	backward
13	1	1	0	1	left
14	1	1	1	0	right
Combinations					
5	0	1	0	1	left forward
6	0	1	1	0	right forward
9	1	0	0	1	left backward
10	1	0	1	0	right backward

5.3 Wired Remote Control with Switches

Previously, only the states of the switches have been read, but now the two drive motors are also to be controlled appropriately. To do this, we include the motor library again and create an instance `motor` of the class `Motor`. With a wired remote control based on push-buttons, it is of course not possible to adjust the speed. We define the PWM value for the speed in the constant V:

```
const int V=255;
```

In the main loop we distinguish between different cases according to the 4-bit value of the push-buttons. The specified program only takes into account the main directions of movement. Combinations (e.g. left forward), in which two contacts are closed simultaneously, currently lead to a standstill. However, the four combined directions of movement can easily be supplemented by additional cases.

```
void loop() {
  int val = ReadButtonState();
  switch (val) {
    // main directions
    case  7: // forward
      motor.write(+V,+V); break;
    case 11: // backward
      motor.write(-V,-V); break;
    case 13: // left
      motor.write(-V,+V); break;
    case 14: // right
      motor.write(+V,-V); break;
    default: // stop
      motor.write(0,0);
  }
}
```

Instead of defining different cases, the PWM values assigned to the respective state value of the buttons for the motor control could also be stored in an array. With 4 bits there are a total of $2^4 = 16$ possible assignments. In addition to the 9 values given in the Tab. 5.1 (8 directions of movement and the rest position, where both switches are in the middle position), 7 assignments remain which are not useful and should not occur (e.g. simultaneously forwards and backwards or left and right, respectively). We mark these with "not defined" and assign the PWM value zero for both motors so that the robot is not moving. For the PWM values, we create the two-dimensional array M of the dimension 16×2. Each line contains a value for the left motor and a value for the right motor. The line number corresponds to the return value of the button's state:

```
const int M[][2] =
  {{ 0, 0}, //  0: not defined
   { 0, 0}, //  1: not defined
   { 0, 0}, //  2: not defined
   { 0, 0}, //  3: not defined
   { 0, 0}, //  4: not defined
   { 0,+V}, //  5: left forward
   {+V, 0}, //  6: right forward
   {+V,+V}, //  7: forward
   { 0, 0}, //  8: not defined
   { 0,-V}, //  9: left backward
   {-V, 0}, // 10: right backward
   {-V,-V}, // 11: backward
   { 0, 0}, // 12: not defined
   {-V,+V}, // 13: left
   {+V,-V}, // 14: right
   { 0, 0}};// 15: stop
```

With this array you can simplify the main loop considerably. After the button is pressed, the required values for the motor control are taken directly from the array M:

```
void loop() {
  int val = ReadButtonState();
  motor.write(M[val][0],M[val][1]);
}
```

Of course, the program can also be extended by an additional LCD output.

5.4 Additional Digital Inputs and Outputs on Arduino Leonardo

The Arduino Leonardo board features the standard 14 digital channels 0,...,13. In addition, the 6 analog channels A0,...,A5 can be used as digital channels 18,...,23 if required. The definition file `pins_arduino.h`, which can be found in the subdirectory `hardware/arduino/avr/variants/leonardo` of the Arduino program folder, provides information about the channels with regard to the numbering [3]:

```
// Map SPI port to 'new' pins D14..D17
static const uint8_t SS   = 17;
static const uint8_t MOSI = 16;
static const uint8_t MISO = 14;
static const uint8_t SCK  = 15;
```

The signals MISO, MOSI and SCK belong to the *In-Circuit Serial Programming (ICSP)* interface, which is provided on (almost) all Arduino boards. The term *In-System Programming (ISP)* is used synonymously [7]. The pin assignment is shown in Fig. 5.7 (left). The special feature of in-system programming is that the microcontroller to be programmed can remain in the system or on the PCB. Data transmission takes place via a serial data bus called *Serial Peripheral Interface (SPI)* [4, 8].

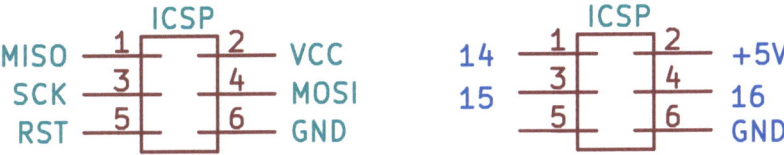

Figure 5.7: Pin assignment of the ICSP interface (left), additional digital channels for the Arduino Leonardo board (right)

This interface could be used to program the controllers of the ATmega series used on Arduino boards [1]. Typically, however, Arduino programming will be done via the USB interface provided on the boards in conjunction with the bootloader. On the Leonardo board, the connectors of the ICSP interface can additionally be used to provide the digital channels 14,...,16, see Fig. 5.7 (right).

The pin Slave Select assigned to digital channel 17 is not part of the ICSP connector, but is related to this interface. The pin is also used to control the RX LED and is not wired out via an external connector. However, there is a through-hole on the PCB for this signal, which can be used to solder a wire or a (single-pole) connector.

Figure 5.8: Additional digital channels on the Arduino Leonardo board

The additional channels can be configured as digital inputs or outputs. For a simple test, the Blink program (File > Examples > 01.Basics > Blink), which is included in the collection of examples, is a good choice. You only need to change the pin used, for example pin 16 for the MOSI connection. After compiling and uploading the program, the level of the respective signal should change every second. This can be verified with an LED and a series resistor as shown in Fig. 5.9. No additional LED is required to check digital channel 17, as this channel is already connected to LED RX on the board.

The Arduino Mirco board is based on the architecture of the Leonardo board. In particular, it uses the same ATmega32u4 processor, which has an integrated USB interface. As a result, the Arduino Micro board, like the Arduino Leonardo

5.4 Additional Digital Inputs and Outputs on Arduino Leonardo

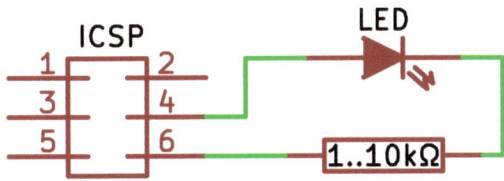

Figure 5.9: Test circuit for the digital channel 16 of the Arduino Leonardo board

board, also has four additional digital channels at its free disposal in the same way as the Arduino Leonardo board. With the Arduino Mirco board, the signal Slave Select is even available, i.e., via a pin on the connector.

Instead of using the ICSP signal lines directly, it would also be conceivable to use SPI to control a suitable peripheral circuit to provide additional ports (a so-called *port expander*), e.g. the IC MCP23S17 [5, 6]. This integrated circuit could be used to operate 16 digital channels, which can be configured as inputs or outputs.

The described port expansion cannot be applied to the Arduino Uno board, although this board also has an ICSP interface (see Section 1.2.1). On the Arduino Uno board, the signal lines of the ICSP interface directly correspond to the digital channels 11 to 13 (and also channel 10 with Slave Select), so that no additional channels are available [4]. In particular, the digital channels 11 to 13 are already occupied by the Arduino Motor shield (see Section 3.2). However, a port extension could be realized via the I²C bus, e.g. with the circuit MCP23017 [6].

References

[1] *Arduino — ArduinoISP.* `https://www.arduino.cc/en/Tutorial/ArduinoISP`.

[2] *Arduino — DigitalPins.* `https://www.arduino.cc/en/Tutorial/DigitalPins`.

[3] *Arduino — Home.* `https://arduino.cc`.

[4] *Arduino — SPI.* `https://www.arduino.cc/en/Reference/SPI`.

[5] *Arduino Playground — MCP23S17 Class for Arduino.* `https://playground.arduino.cc/Main/MCP23S17`.

[6] Microchip: *MCP23017/MCP23S17, 16-Bit I/O Expander with Serial Interface*, 2007. Datasheet.

[7] Wikipedia contributors: *In-system programming — Wikipedia, the free encyclopedia.* `https://en.wikipedia.org/w/index.php?title=In-system_programming`.

[8] Wikipedia contributors: *Serial Peripheral Interface — Wikipedia, the free encyclopedia.* `https://en.wikipedia.org/w/index.php?title=Serial_Peripheral_Interface`.

Chapter 6

Measurement of Analog Signals

6.1 Connecting a Potentiometer

All considered Arduino boards have the analog input channels A0 to A5. (The Leonardo board has further analog channels available, but for their use you have to do without the corresponding digital input/output channels.) In combination with the Arduino motor shield R3, it is expected that A0 and A1 are intended for the measurement of the motor current and are thus occupied. Hence, for a first experiment the channel A2 is used, to which a potentiometer is connected (see Fig. 6.1).

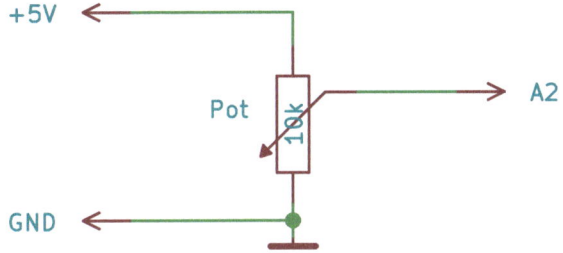

Figure 6.1: Wiring of a potentiometer

To read an analog channel, two easy-to-understand examples can be found in the Arduino IDE under File > Examples > 03.Analog. In the example program

`AnalogInOutSerial`, the read values are sent to the terminal via the serial interface. With `AnalogInput`, the flashing frequency of an LED is controlled by the read voltage value.

We want to display the measured values on the LCD panel. First of all, you have to prepare the program for the operation of the LCD module:

```
#include <LiquidCrystal.h>
LiquidCrystal lcd(10, 2, 7, 6, 5, 4);
```

The analog channel A2 is assigned to the constant `Pot`. For the value to be read out from the *analog-to-digital converter (ADC)*, the integer variable `val` is used.

```
const byte Pot = A2;
int val;
```

The ADC maps the voltage range from 0 V to 5 V on (integer) values from 0 to 1023. With the following conversion factor you can determine the corresponding voltage for the analog value read out:

```
const float scale=5.0/1023;
```

In the function `setup`, an initialization of the analog channels is not necessary, but for the display on the LCD module:

```
void setup() {
  lcd.begin(20, 4);
  lcd.print("Potentiometer:");
}
```

For the actual reading of the ADC values, the function `analogRead` is available in the Arduino environment. The used analog channel has to be given as an argument:

```
val = analogRead(Pot);
```

This reading is cyclically carried out in the main loop. The second line of the LCD screen shows the read out value of the ADC, and in the third line the calculated voltage using the above mentioned conversion factor is shown:

6.1 Connecting a Potentiometer

```
void loop() {
  val = analogRead(Pot);
  // ADC value
  lcd.setCursor(0, 1);
  lcd.print("ADC: ");
  lcd.print(val);
  lcd.print("   "); // three spaces
  // voltage
  lcd.setCursor(0, 2);
  lcd.print("Voltage: ");
  lcd.print(scale*val);
  lcd.print(" V ");
}
```

The ADC value is printed left-justified and therefore has a different length as a character string depending on the value to be displayed. For example, if only a three-digit value is shown after a four-digit number has been displayed, the last digit of the four-digit number would remain. To delete these "remaining" digits, three spaces are printed at the end.

Overall, the output with several **print** commands is comparatively cumbersome. If the library **Streaming** [2] is also included, the main loop is simplified noticeably:

```
void loop() {
  val = analogRead(Pot);
  lcd.setCursor(0, 1);
  lcd<<"ADC: "<<val<<"   ";
  lcd.setCursor(0, 2);
  lcd<<"Voltage: "<<scale*val<<" V ";
}
```

The conversion factor **scale** is a floating point number. The microcontrollers of the ATmega series do not have a hardware floating point unit, the corresponding operations are simulated by software. This costs memory space and computation time. One possible alternative would be to display the voltage not as a floating point number in volts (V) but as an integer value in millivolts (mV). In addition to the existing integer variable **val**, a further integer variable **v** is declared for the voltage in mV:

```
int val, v;
```

Furthermore, the ADC values ranging from 0 to 1023 must be scaled to 0 to 5000 for the display in mV. The function `map` is available in the Arduino environment for this purpose:

```
v = map (val, 0, 1023, 0, 5000);
```

The following main loop could be used for the voltage output in millivolts:

```
void loop() {
  val = analogRead(Pot);
  lcd.setCursor(0, 1);
  lcd<<"ADC: "<<val<<"    ";
  lcd.setCursor(0, 2);
  v = map (val, 0, 1023, 0, 5000);
  lcd<<"Voltage: "<<v<<" mV   ";
}
```

The integer value stored in the variable v can also be displayed as a decimal number in V with reasonable effort. To do this, you include the standard library `stdio.h` and create two string variables:

```
char line1[21];
char line2[21];
```

The main loop could then look like this:

```
void loop() {
  val = analogRead(Pot);
  lcd.setCursor(0, 1);
  sprintf(line1, "ADC: %5d", val);
  lcd.print(line1);
  lcd.setCursor(0, 2);
  v = map (Wert, 0, 1023, 0, 5000);
  sprintf(line2, "Voltage: %1d.%03d", v/1000, v%1000);
  lcd.print(line2);
}
```

With "%5d" or "%1d" an integer variable with 5 digits or with 1 digit is displayed. The format string "%03d" generates a 3-digit output in which the leading digits are filled with zeros. The voltage component in V is determined with the integer division `v/1000`, while the part determined via the division remainder `v%1000` gives the digits after the decimal point specified in mV.

6.2 Wired Control via two Sliding Potentiometers

With two potentiometers you can easily set up a wired remote control. The two potentiometers are connected to the robot via a four-core flexible cable. The middle connections of the potentiometers must be connected to two analog inputs of the Arduino board. For the circuit diagram shown in Fig. 6.2, channels A2 and A3 are used.

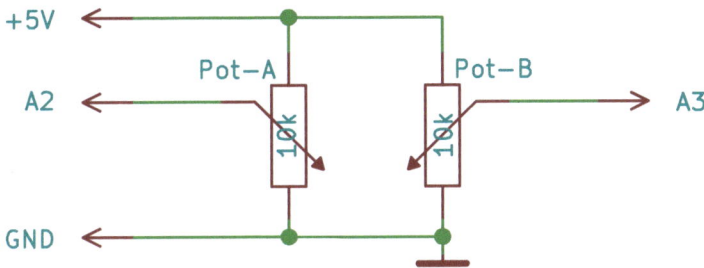

Figure 6.2: Wiring of the potentiometers

Each of the two potentiometers is used to adjust the direction of rotation or angular speed for a motor. The potentiometer A is for the left motor and the potentiometer B for the right motor (see Fig. 6.3).

On the software side, the required libraries are included first:

```
// libraries
#include <LiquidCrystal.h>
#include <Motor.h>
#include <stdio.h>
```

An instance is created for each of the C++ classes for LCD and motor control:

```
// create objects
LiquidCrystal lcd(10, 2, 7, 6, 5, 4);
Motor motor(12, 13, 3, 11);
```

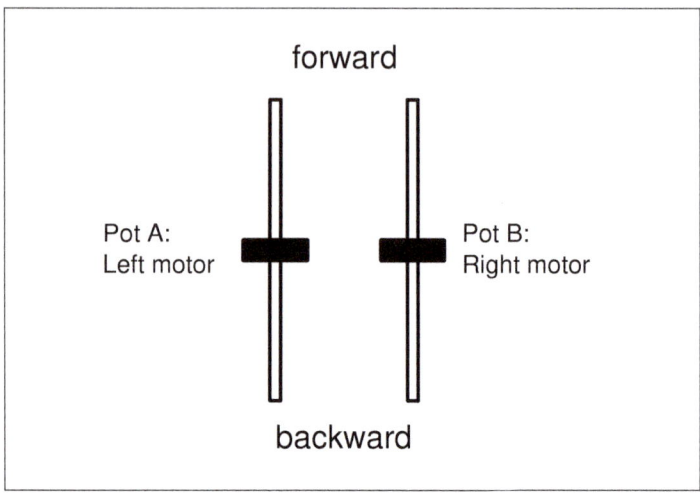

Figure 6.3: Remote control with two potentiometers

The analog channels A2 and A3 used for the two potentiometers are defined as constants:

```
// pins for pots
const byte PotA = A2; // left
const byte PotB = A3; // right
```

We declare integer variables for the ADC values of the potentiometers and the output values for the motors to be calculated from them:

```
// ADC signals
int valA, valB;
// motor signals
int driveA, driveB;
```

In the function `setup`, the LCD display and the class `Motor` are initialized:

```
void setup() {
  lcd.begin(20, 4);
  lcd.print("Remote control: Pot");
  motor.begin();
}
```

For both channels the read out analog values of the potentiometers as well as the output values for the motors should be displayed. Since the output of

6.3 Current Measurement on the Arduino Motor Shield

these values is almost the same for both channels, a function is created for the output. The channels are selected via the variable **no**, which has the value 0 for channel A and the value 1 for channel B.

```
// LCD output
void output (short no, int val, int drive) {
  char line[21];
  lcd.setCursor (0, 1+no);
  sprintf(line, "Channel %c: %4u / %+3d", 'A'+no, val, drive);
  lcd.print(line);
}
```

In the main loop, the potentiometer values (0...1023) are first read and then scaled to the value range of -200...200 for controlling the motors. For monitoring purposes, these values are displayed on the LCD. Then the motors are controlled. With a small time delay, too hectic robot reactions can be avoided:

```
void loop() {
  // ADC reading
  valA = analogRead(PotA);
  valB = analogRead(PotB);
  // scaling
  driveA = map (valA, 0, 1023, -200, 200);
  driveB = map (valB, 0, 1023, -200, 200);
  // LCD output
  output(0, valA, driveA);
  output(1, valB, driveB);
  // motor control
  motor.write (driveA, driveB);
  delay(100);
}
```

6.3 Current Measurement on the Arduino Motor Shield

The official Arduino motor shield R3 has the ability to measure motor currents. The analog input A0 is available for motor A, input A1 for motor B. The motor current measurement is explained for motor A, the same procedure also applies to motor B. To verify the measurement result, a multimeter operated as amperemeter should be connected in series to motor A for the tests.

The motor current should be shown on the LCD screen. To do this, you include the corresponding libraries and create the instance `lcd` of the class `LiquidCrystal`:

```
#include <LiquidCrystal.h>
#include <Streaming.h>
LiquidCrystal lcd(10, 2, 7, 6, 5, 4);
```

The motor control should be done here with the native functions of the Arduino environment, i.e., without the library `Motor`. The addresses are therefore defined for both motor control as well as current measurement:

```
const byte DIRA = 12;
const byte PWMA =  3;
const byte CURA = A0;
```

In the function `setup`, the LCD is initialized in the usual way, the digital channels used for motor control are declared as outputs and motor A is set into a suitable motion (here: forward motion with PWM value 200):

```
void setup() {
  lcd.begin(20, 4);
  lcd.print("Motor current:");
  // motor signals
  pinMode (DIRA, OUTPUT);
  pinMode (PWMA, OUTPUT);
  // switch on motor A
  digitalWrite (DIRA, LOW);
  analogWrite (PWMA, 200);
}
```

The value read by the ADC is an integer, the current could be interpreted as a floating point number. For efficiency reasons, however, the current should also be represented by an integer number. For this purpose we declare two variables:

```
int val;
int cur;
```

The analog channel for current measurement is read in the same way as the potentiometer values considered at the beginning of this chapter:

6.3 Current Measurement on the Arduino Motor Shield

```
val = analogRead(CURA);
```

With the Arduino motor shield the maximum allowed current of 2 A is mapped to the voltage of 3.3 V [1]. This corresponds to 1.65 V/A. The ADC maps the voltage range from 0 V to 5 V to values from 0 to 1023, which would result in a current from 0 mA to 3030 mA = 3.03 A across this range. The conversion of the range from 0 to 1023 to the current from 0 to 3030 given in mA can again be done with the function `map`:

```
cur = map(val, 0, 1023, 0, 3030);
```

The main loop for current measurement and visualization is shown below. In addition, each cycle is delayed by 500 ms so that the display does not flicker to much.

```
void loop() {
  val = analogRead(CURA);
  lcd.setCursor(0, 1);
  lcd<<"ADC: "<<val<<" ";
  // current in mA
  cur = map(val, 0, 1023, 0, 3030);
  lcd.setCursor(0, 2);
  lcd<<"Current: "<<cur<<" mA ";
  delay(500);
}
```

In the experiment, the measured values varied significantly from one sampling step to the next. Here it needs to be clarified that with a DC motor the current is cyclically disrupted by the commutator. During these switching operations, voltage peaks can also be caused by the winding inductances. A low-pass filter can help here, with which the measured values can be smoothed. Before the value `val` is reassigned, the arithmetic mean between the old value and the actual measured value is calculated on the right-hand side:

```
val = (val + analogRead(CURA)) / 2;
```

With this filtering, the displayed value behaves noticeably less erratic. If the wheel is held briefly, a noticeably higher current should flow.

6.4 Obstacle Avoidance by Means of Motor Current Measurement

Push-buttons or distance sensors are normally used for obstacle detection. However, an obstacle can also be detected via the motor current. If the robot is stopped or braked, the motor current increases noticeably. This effect is used in this section to allow obstacle avoidance without additional sensors. The measured values are solely obtained using the Arduino motor shield, which provides two channels for motor current measurement.

First, the libraries for the LCD module and the motor control are included as usual:

```
// libraries
#include <LiquidCrystal.h>
#include <Motor.h>
```

The C++ classes provided by this are instantiated according to the wiring of LCD and motor:

```
// create objects
LiquidCrystal lcd(10, 2, 7, 6, 5, 4);
Motor motor(12, 13, 3, 11);
```

The Ardunio motor shield uses the analog channels A0 and A1 to measure the motor currents:

```
// pins for current measurement
const byte CURA = A0;
const byte CURB = A1;
```

When driving straight ahead without obstacles, the motor current ADC values were in the range 20...45 (after filtering), while values of 100 and more occurred when blocking the wheels. The value 70 for the constant `Imax` is in between and used to distinguish between these two situations.

```
// max value motor current
const byte Imax = 70;
```

The ADC values for the motor currents are stored in the variables `iA` or `iB`. The declaration of these auxiliary variables is necessary because of the filtering.

6.4 Obstacle Avoidance by Means of Motor Current Measurement

```
// motor currents
int iA = 0;
int iB = 0;
```

Additional constants are required to specify the evasive maneuvers. The value `Speed` specifies the PWM value for the motor control. Values up to 255 are possible here. The value `T_back` indicates the time in ms for backwards driving (after an obstacle has been detected). The time required for a lateral rotation of approx. 90° (again in ms) is stored in `T_rot`:

```
// constants for evasive maneuvers
const int Speed  = 180;
const int T_back = 800; // time interval backwards driving
const int T_rot  = 500; // time interval lateral rotation
```

The initialization of the classes for LCD and motor control is carried out as usual in the function `setup`:

```
void setup() {
  lcd.begin(20, 4);
  motor.begin();
}
```

Next is the main loop:

```
void loop()
{
  ...
}
```

In the main loop, both analog channels are read and filtered by averaging:

```
  iA = (iA + analogRead(CURA)) / 2;
  iB = (iB + analogRead(CURB)) / 2;
```

The values read by the ADCs should be displayed in the first line of the LCD panel:

```
  lcd.home();
  lcd.print("Ia: ");
  lcd.print(iA);
  lcd.print(" Ib: ");
```

```
lcd.print(iB);
lcd.print(" ");
```

If both (filtered) values for the measured motor currents do not exceed the specified limit `Imax`, the robot is not affected in its movement and can continue to move forward.

```
if (max(iA, iB)<=Imax) {
  // forward
  lcd.setCursor(0,1);
  lcd.print("Forward    ");
  motor.write (Speed, Speed);
  delay(100);
}
```

Otherwise, if at least one value for the motor current is above the threshold `Imax`, the movement of the robot is impaired. This case is classified as an obstacle. The robot first moves backwards for a time and then turns to the side. The motor is briefly switched off for the next run so that the change of direction does not affect the next current measurement.

```
else {
  // backward
  lcd.setCursor(0,1);
  lcd.print("Backward    ");
  motor.write (-Speed,-Speed);
  delay(T_back);
  lcd.setCursor(0,1);
  lcd.print("To the right");
  motor.write (Speed,-Speed);
  delay(T_rot);
  motor.write (0,0);
  delay(100);
  iA=0;
  iB=0;
}
```

6.5 Motor Voltage Measurement on the Velleman Motor Shield

The motor shield from Velleman has the option of measuring the motor voltage. Depending on the position of jumper SK7 the motor is supplied either by the supply voltage V_in of the Arduino board or by an external voltage V_ext. In the following, the motor supply voltage is referred to as V_Power. Resistors R9 and R10 and electrolytic capacitor C3 are responsible for generating the measurement signal on the motor shield (see [3, 4] and Fig. 6.4).

Figure 6.4: Velleman motor shield with marked components R9, R10 and C3

The supply voltage V_Power is directed via the voltage divider consisting of R9 and R10 to the analog input A5, whose voltage is denoted by V_ADC. The electrolytic capacitor C3 is also provided for smoothing as shown in Fig. 6.5. The following voltage drop occurs at the voltage divider's output:

$$\frac{V_\text{ADC}}{V_\text{Power}} = \frac{R10}{R9 + R10} = \frac{10\,\text{k}\Omega}{110\,\text{k}\Omega} = \frac{1}{11}.$$

The voltage V_ADC applied to the ADC is therefore factor 11 lower than the supply voltage V_Power. In other words: Multiplying the voltage determined at the ADC by a factor of 11 results in the supply voltage. The measurement of

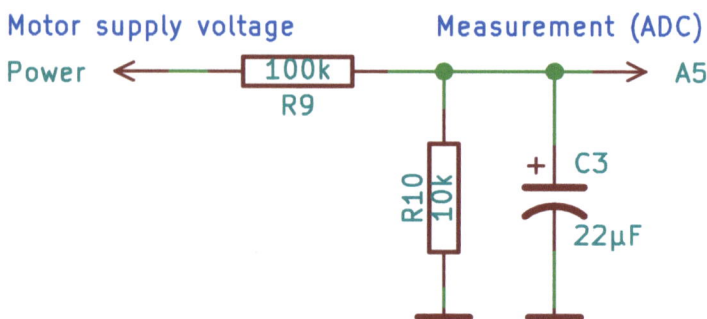

Figure 6.5: Circuit diagram for measuring the motor supply voltage of the Velleman motor shield [3, 4]

the supply voltage on the motor side would therefore be essentially the same as the voltage measurement on the potentiometer described at the beginning of this chapter. Instead of the conversion factor used for the potentiometer

```
const float scale=5.0/1023;
```

you would now use a conversion factor increased by a factor of 11:

```
const float scale=55.0/1023;
```

References

[1] *Arduino Motor Shield.* https://arduino.cc/en/Main/ArduinoMotorShieldR3.

[2] *Streaming | Arduiniana.* http://arduiniana.org/libraries/streaming/.

[3] Velleman: *KA03 Motor & Power shield Arduino®.* Illustrated assembly manual.

[4] Velleman: *VMA03 Motor & Power shield Arduino®.* Manual.

Chapter 7

Distance Measurement

7.1 Distance Sensors based on Optical Triangulation

With a rangefinder or distance sensor, the robot is able to detect and avoid obstacles before a contact or collision. There are numerous measuring principles and sensor types for distance measurement. One way to determine the distance is optical triangulation (see Fig. 7.1). A light beam is emitted which is reflected by the object to be measured. This reflection is perceived via a light-sensitive element. Typically, the reflected light beam is directed via a lens or prism onto a CCD line sensor. Depending on the distance, the angle of the reflected light beam and thus the position on the CCD line changes. The CCD sensor is a light-sensitive integrated component, where the abbreviation CCD stands for *charge coupled device*. In digital cameras, CCD sensors with a matrix-like (two-dimensional) arrangement of the individual photo elements are used, whereas a line-like (one-dimensional) arrangement is sufficient for distance measurement.

Sharp distance sensors are very popular in mobile robotics. The types specified in Tab. 7.1 are commonly used to measure distances in the range of a few decimetres. The GP2Y0A21YK0F and GP2Y0A41SK0F are virtually indistinguishable with regard to their design (see Fig. 7.2). The GP2Y0A02YK0F sensor has a similar design but is slightly larger [5]. Fig. 7.3 shows one of the above sensors, the GP2Y0A41SK0F, on the front of the mobile robot.

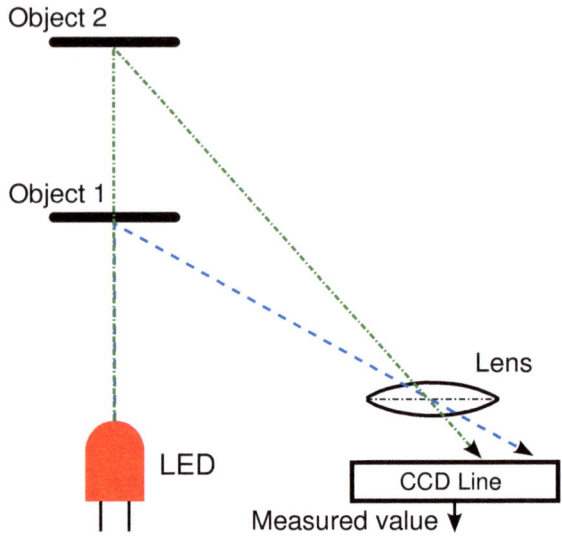

Figure 7.1: Distance measurement by optical triangulation

Table 7.1: Popular distance sensors from Sharp [5, 6, 7]

New type	Old type	Range of measurement	
Sharp GP2Y0A41SK0F	Sharp GP2D120	4...30 cm	(1.5...12 in)
Sharp GP2Y0A21YK0F	Sharp GP2D12	10...80 cm	(4...31 in)
Sharp GP2Y0A02YK0F	—	20...150 cm	(8...59 in)

Figure 7.2: Sharp distance sensors GP2Y0A21YK0F (left) and GP2Y0A41SK0F (right)

7.1 Distance Sensors based on Optical Triangulation   77

Figure 7.3: Sharp distance sensor in the front area of the mobile robot

Wiring is very simple. The sensor has three pins. Two of these pins are used for power supply (ground and +5 V), the third pin supplies the measurement signal as analog voltage value. For this signal, the analog input A2 of the Arduino board was used (Fig. 7.4).

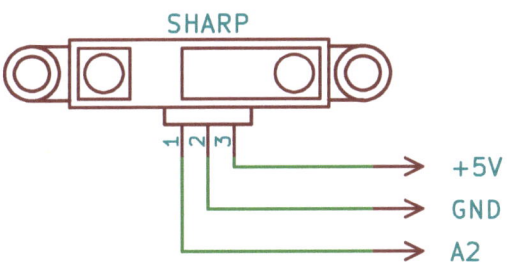

Figure 7.4: Wiring of a Sharp distance sensor

For the distance sensor reading on the software side, we of course need the libraries relevant for the LCD output:

```
#include <LiquidCrystal.h>
#include <stdio.h>

LiquidCrystal lcd(10, 2, 7, 6, 5, 4);
```

Next, we define the analog channel (pin) for the reading of the sensor output and create the variable x for the obtained value:

```
const byte pinSensor = A2;
unsigned x=0;
char line[21];
```

In the function **setup** only the LCD module must be initialized:

```
void setup() {
  lcd.begin(20, 4);
}
```

The actual sensor reading is done with the function **analogRead**. In addition, the sensor signal is filtered. Only the filtered ADC value is displayed, not the distance. Since the output voltage of the sensor is in the range of about $0\ldots3.1$ V, only about 2/3 of the ADC measuring range is used. The displayed ADC values should therefore be in the range $0\ldots635$. To prevent the LCD output from flickering too much, a waiting time of 200 ms was inserted for each pass of the main loop:

```
void loop()
{
  // measurment and filtering
  x = (4*x + analogRead(pinSensor)) / 5;
  // output
  lcd.home();
  sprintf(line, "ADC: %4u", x);
  lcd.print(line);
  delay(200);
}
```

For the setup of the distance sensor, the robot can remain connected to the PC. In this case, the signal values could be transferred to the PC via the USB cable and the LCD would not be needed. Then, the libraries **LiquidCrystal.h** and **stdio.h** would no longer be necessary. In the function **setup** you would initialize the serial connection instead of the LCD:

```
void setup()
{
  Serial.begin(9600);
}
```

In the main loop, only the output needs to be adjusted:

```
void loop()
{
  x = (4*x + analogRead(pinSensor)) / 5;
  Serial.println(x);
  delay(200);
}
```

The measured values can be displayed with the Arduino IDE under Tools > Serial Monitor.

7.2 Calibration of the Sensor

The values recorded by the ADC from the distance sensor can also be converted into standard length dimensions, e.g. mm, cm or inch (in). For this purpose, either the characteristic curves specified in the data sheets can be used or you have to carry out your own measurements. In the latter case, the program described in the previous section can be used to display the ADC values. In Tab. 7.2, $N = 6$ measured values are listed. These measured values are additionally shown as black dots in Fig. 7.5. For further processing, these measured values were numbered consecutively for $i = 0, \ldots, N - 1$.

Table 7.2: Measured values for calibrating the Sharp GP2Y0A41SK0F distance sensor

Number i	Distance l_i in mm	ADC value x_i
0	50	488
1	100	250
2	150	166
3	200	118
4	250	95
5	300	83

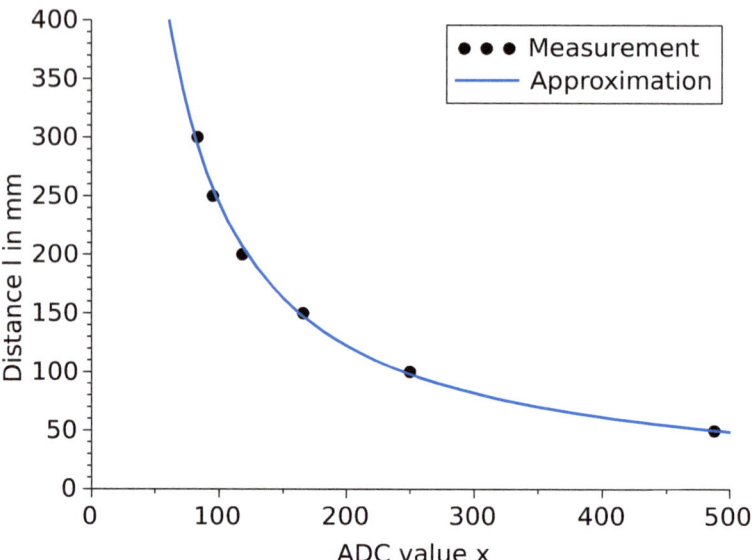

Figure 7.5: Measured values and approximated characteristic of the Sharp GP2Y0A41SK0F sensor

The measured values shown in the diagram clearly do not lie on a straight line. Therefore, a linear approach to describe the characteristic curve does not make sense. However, the measured values could approximately be described by a hyperbole, so that the following approach would be conceivable at first:

$$l = \frac{K}{x}$$

A value x supplied by the ADC is in the range of $0\ldots 1023$, which could also be zero. To avoid this singularity (division by zero), we modify the approach as follows:

$$l = \frac{K}{x+1}$$

The coefficient K can then be calculated with the following formula:

$$K = \left(\sum_{i=0}^{N-1} \frac{l_i}{x_i+1}\right) / \left(\sum_{i=0}^{N-1} \frac{1}{(x_i+1)^2}\right)$$

The coefficient $K \approx 24565$ was determined for the measured values shown

7.2 Calibration of the Sensor

in Tab. 7.2. The characteristic determined with this coefficient (blue hyperbole in Fig. 7.5) shows an extraordinarily good agreement with the measured values.

If you use your own measured values or a Sharp sensor for another measuring range, you must evaluate the above formula for calculating the coefficient K yourself. This calculation can also be done on the Arduino board. The first step is to store the measured values in suitable data fields:

```
const unsigned N=6; // number of measured values
int L[N] = { 50, 100, 150, 200, 250, 300};
int X[N] = {488, 250, 166, 118, 95, 83};
```

A global variable is declared for the coefficient K:

```
// coefficient
double K;
```

The actual calculation according to the above formula only needs to be performed once and is therefore done in the function `setup`. For the numerator there is the auxiliary variable b, for the denominator the auxiliary variable a. The serial interface is initialized for output:

```
void setup () {
  // calculation
  double a=0, b=0, c;
  for (int i=0; i<N; i++) {
    c=X[i]+1.0;
    a+=1/(c*c);
    b+=L[i]/c;
  }
  K=b/a;
  // initialize output
  Serial.begin(9600);
}
```

The actual output takes place in the main loop. The value of the coefficient K is displayed once a second.

```
void loop () {
  Serial.println(K);
  delay(1000);
}
```

With a new program, the robot could be converted into a distance meter. The coefficient K would be stored as a constant in the program:

```
// constant for hyperbola
const unsigned K = 24565;
```

In addition to the variable x for the sensor value, a variable l for the distance in mm would also have to be declared:

```
unsigned l,x=0;
```

In the main loop, the sensor is read and filtered slightly less strongly than in the previous program. Using the above formula, the filtered ADC value x is then converted by to the length l:

```
void loop() {
  x = (2*x + analogRead(pinSensor)) / 3;
  l = K / (x+1);
  lcd.home();
  sprintf(line, "L in mm: %5u", l);
  lcd.print(line);
  delay(200);
}
```

7.3 Obstacle Avoidance with one Distance Sensor

In this section, the mobile robot is programmed so that it detects obstacles in the direction of travel and avoids them laterally if necessary. The libraries for LCD and motor control are included in the usual way. The streaming library is additionally provided:

```
// libraries
#include <LiquidCrystal.h>
#include <Streaming.h>
#include <Motor.h>
```

The LCD is also set up as before:

```
// create object
LiquidCrystal lcd(10, 2, 7, 6, 5, 4);
```

7.3 Obstacle Avoidance with one Distance Sensor

With regard to motor control, the class `MotorNew` defines additional submaneuvers. In particular, the call of the basic types of movement (forward, backward, right and left) is to be simplified. The four additional functions can be called either without an argument or with exactly one argument. When called with an argument of type `byte`, the corresponding velocity value is specified in the range 0...255. If the respective functions are called without an argument, the velocity value is passed via the constant `Vmax` defined in the base class `Motor` (typically 255).

```
class MotorNew : public Motor {
  // additional submaneuvers
  public:
  void forward  (byte speed=Vmax) {write (+speed,+speed); };
  void backward (byte speed=Vmax) {write (-speed,-speed); };
  void right    (byte speed=Vmax) {write (+speed,-speed); };
  void left     (byte speed=Vmax) {write (-speed,+speed); };
};
```

During instantiation, the standard constructor of the class `MotorNew` is called, which in turn calls the standard constructor of the base class `Motor`. The standard constructor sets the pin assignment for the Arduino motor shield.

```
MotorNew motor;
```

In case of a different pin assignment you would have to define an additional constructor for the class `MotorNew`, which calls the non-standard constructor of the base class `Motor`. Note that although the derived class adopts the methods of the basis class, the constructors are not inherited and must therefore be created again if necessary.

However, the function `setup` has no special features:

```
void setup() {
  lcd.begin(20, 4);
  motor.begin();
}
```

The distance sensor is again connected to the analogue channel A2. With the sensor type used, the ADC value `xMax=200` corresponds to a distance in the range of 10...15 cm (i.e., 4...6 in). The ADC value is stored in the integer variable x.

```
const byte pinSensor = A2;
const int  xMax = 200;
int x;
```

In the main loop, movements are to be carried out at maximum travel speed (PWM value 255). A 90° rotation requires approx. 400 ms for the robot used.

```
// constants for evasive maneuvers
const int T_back = 800; // time intervall backwards driving
const int T_rot  = 400; // time intervall lateral rotation
```

All in all, a very straightforward main loop is obtained, whereby the sensor value is not filtered:

```
void loop() {
  x = analogRead(pinSensor);
  lcd.home();
  lcd<<"ADC: "<<x<<" ";
  if (x <= xMax) {
    // forward
    motor.forward();
  }
  else {
    // backward
    motor.backward();
    delay(T_back);
    // rotate
    motor.right();
    delay(T_rot);
  }
  delay(100);
}
```

7.4 Capturing the Surroundings and Searching for a new Direction

With a single distance sensor, the robot can detect its entire environment through a 360° rotation. In this section, a program is developed to move the

7.4 Capturing the Surroundings and Searching for a new Direction

robot in the direction where obstacles are furthest away as sketched in Fig. 7.6.

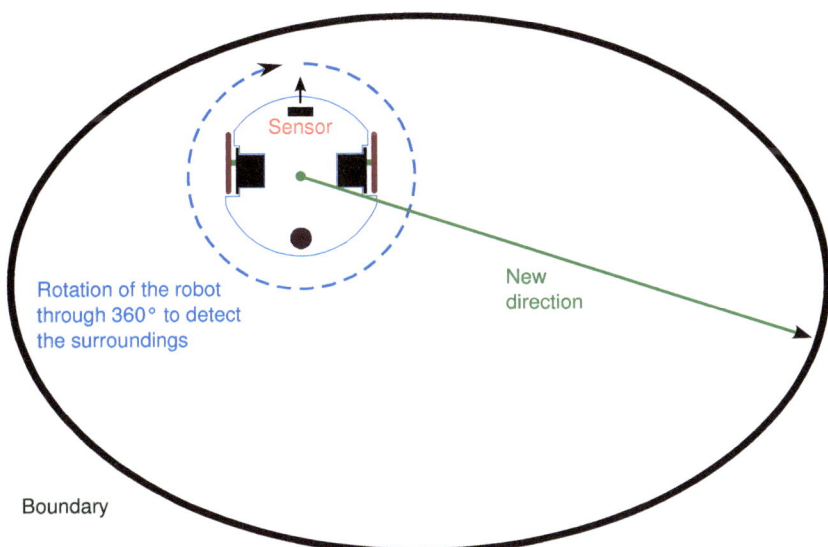

Figure 7.6: Determination of the new direction of travel

Let's start with the preparation. The motor library must be included, if necessary also the library for the LCD. The class `MotorNew` derived from the basic class `Motor`, which we presented in the last section, is also used. This class is extended by the method `stop`, which stops both motors:

```
class MotorNew : public Motor {
  // submaneuvers
  public:
  void forward  (byte speed=Vmax) {write (+speed,+speed); };
  void backward (byte speed=Vmax) {write (-speed,-speed); };
  void right    (byte speed=Vmax) {write (+speed,-speed); };
  void left     (byte speed=Vmax) {write (-speed,+speed); };
  void stop ()   {write (0,0); };
};
```

As an extension of the implementation given here, it would also be conceivable to use the braking function provided in the Arduino motor shield. Regardless of this, the instantiation is carried out in the usual way:

```
MotorNew motor;
```

The distance sensor is again wired to the analog input A2:

```
const byte pinSensor = A2;
```

The next parameters describe the 360° rotation. A measured value for the distance is recorded at `N` places after the waiting time stored in `T_step` (here: 10 ms) and written into an element of the one-dimensional integer array `X`.

```
const int T_step = 10;
const unsigned N = 180;
unsigned X[N];
```

The following C function captures the distance information for the robot's environment. First, a clockwise rotation of the robot is triggered. During rotation, the sensor is read out cyclically and the measured value is stored in the data array. After the waiting time defined in `T_step`, the next value is processed. After `N` steps, the motors are stopped and rotation is finished.

```
// capture surroundings
void capture() {
  motor.right();
  for (int i=0; i<N; i++) {
    X[i]=analogRead(pinSensor);
    delay(T_step);
  }
  motor.stop();
}
```

The constant `N` must be selected as a function of the motor speed and the waiting time `T_step` such that rotation to 360° is completed. By calling `motor.right()` without an argument, the default value `Vmax` is used; if necessary, the motor speed can also be adjusted. The values $N = 180$ used here means that the full circle is recorded in steps of 2° each: $360°/N = 360°/180 = 2°$.

Next, the measured values stored in array `X` are evaluated. The maximum distance to an obstacle corresponds to the smallest value read in for Sharp distance sensors. After passing through the `for` loop, this value is stored in the variable `xmin`, the corresponding field entry in `imin`. The function `evaluation` returns the array index for the new direction.

7.4 Capturing the Surroundings and Searching for a new Direction

```c
// determine the new direction
int evaluation() {
  unsigned imin=0, xmin=1024;
  for (int i=0; i<N; i++) {
    if (X[i]<xmin) { imin=i; xmin=X[i]; }
  }
  return imin;
}
```

Now we set the calculated direction by turning. The robot should then move in the new direction. This travel time is defined with the following constant. This time interval should not be too large, because otherwise the robot leaves the captured area during this maneuver.

```c
// time interval for driving
const int T_forward=500;
```

The actual driving maneuver is implemented in the function **move**. The function argument i from 0 to $N-1$ indicates the segment number for the full circle divided into N equal parts. In our case, the angle to be set would be the ith multiple of 2°. The angle is set by turning the robot clockwise by i time units (**T_step**). This is followed by the journey in the new direction for the time interval **T_forward**.

```c
// move into new direction
void move (int i) {
  // rotate
  motor.right();
  delay(i*T_step);
  // move
  motor.forward();
  delay(T_forward);
  motor.stop();
}
```

In the main loop, the three C functions **capture**, **evaluation** and **move** are called sequentially. In addition, there is a short output on the LCD.

```
void loop() {
  unsigned imin;
  lcd.clear();
  lcd.print("Read values ...");
  // capture surroundings
  capture();
  // determine the new direction
  imin=evaluation();
  lcd.clear();
  lcd<<"i= "<<imin;
  // rotate and move into new direction
  move(imin);
}
```

At the beginning, the robot must be positioned on the playing field such that the minimum distance specified for the sensor used is guaranteed (4 cm for the Sharp GP2Y0A41SK0F or 10 cm for the Sharp GP2Y0A21YK0F). If one does not keep this minimum distance at the beginning, then one gets a measuring signal, which suggests a very large distance. The robot would then incorrectly drive against the boundary of the playing field and not in the direction in which there would actually be enough room.

The described implementation can be improved in several ways. The new direction of travel is always set by turning it clockwise. For angles exceeding 180°, the desired angle can be reached more quickly by turning to the left. To do this, you could change the routine **move** as follows:

```
// move into new direction
void move (int i) {
  // direction of rotation
  if (2*i<N) {
    motor.right();
    delay(i*T_step);
  }
  else {
    motor.left();
    delay((N-i)*T_step);
  }
  // move
```

7.4 Capturing the Surroundings and Searching for a new Direction

```
  motor.forward();
  delay(T_forward);
  motor.stop();
}
```

In the above mentioned implementation of the routine `evaluation` it was assumed that the playing field is completely bordered. With a small gap in the border, the sensor would detect a large distance and the routine `evaluation` could select this as a new direction (see Fig. 7.7).

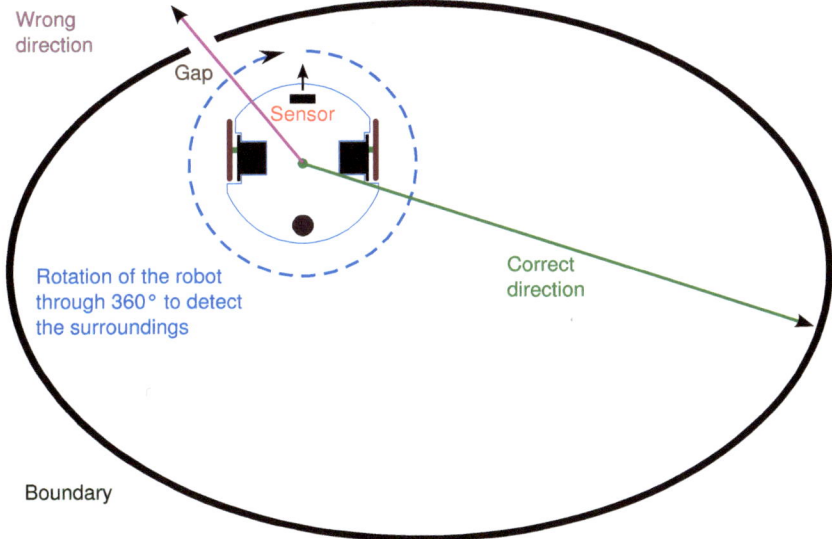

Figure 7.7: Problem in the determination of the new direction of travel

This problem can be avoided by not only incorporating the measured value in the ith step but also taking an entire angular segment into account when determining the new direction. The 20 previous and 20 subsequent field entries were used for the improved implementation. If the field index $i + j$ with $-20 \leq j \leq 20$ is below 0 or above $N-1$, the field number is corrected by modulo calculation (division with remainder):

```
// determine the new direction
int evaluation() {
  unsigned imin=0, xmin=1024, jmax;
  for (int i=0; i<N; i++) {      // angle
```

```
    jmax=0;
    for (int j=-20; j<=20; j++)    // angular segment
      jmax=max(jmax,X[(i+j) % N]);
    if (jmax<xmin) {                // gap?
      imin=i;
      xmin=jmax;
    }
  }
  return imin;
}
```

The functions `capture`, `evaluation` and `move` could also be combined in one class together with the data array `X`. If you also declare these routines as *virtual functions*, you could easily implement improvements in derived classes.

7.5 Obstacle Avoidance with two Distance Sensors

A common design variant in mobile robotics is the mounting of two distance sensors on the front of the robot. Fig. 7.8 shows the mobile robot with two Sharp distance sensors.

Figure 7.8: Two Sharp distance sensors in the front area of the mobile robot

7.5 Obstacle Avoidance with two Distance Sensors

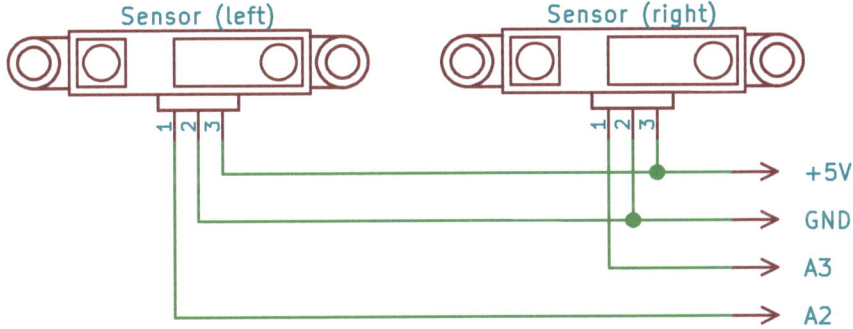

Figure 7.9: Wiring of two Sharp distance sensors

The sensors are supplied with +5 V and can be read via the analog inputs A2 and A3. Channel A2 is for the left sensor, channel A3 for the right sensor (Fig. 7.9). On the software side, the analog inputs are again defined as constants:

```
const byte pinSensorA = A2; // left
const byte pinSensorB = A3; // right
```

As with the obstacle avoidance with one distance sensor, the constant xMax defines the sensor value from which an obstacle is detected.

```
// distance: 10 .. 15 cm
const int xMax = 200;
```

Both sensors are read by the C function readSensor. If this function returns the value 0, none of the sensors detected a near obstacle. With the return values 1 or 2, an obstacle was detected either on the left sensor or on the right sensor respectively. If both sensors detect an obstacle, the value 3 is returned.

```
int readSensor() {
  int x;
  x = analogRead(A2)<xMax ? 0 : 1;
  x+= analogRead(A3)<xMax ? 0 : 2;
  return x;
}
```

For the visual output of the sensor reading, the four options are stored in text form as an array:

```
char *Message[] = {"None", "Left", "Right", "Straight ahead"};
```

The corresponding text output for obstacle detection or classification takes place in the main loop, so that the functionality of the sensors can be tested:

```
void loop() {
  int x=readSensor();
  lcd.clear();
  lcd.print("Obstracle:");
  lcd.setCursor(0,1);
  lcd.print(Message[x]);
  delay(200);
}
```

After testing the sensors, we can move on to implementing the evasive maneuver. Of course, the motors of the robot must be controlled. The respective direction of travel was previously passed either directly as PWM values in the method `write` of the class `Motor` or as one of the methods `forward`, `backward`, `right` or `left` of the class `MotorNew` dealt with in the last sections. Another way is to define required standard directions as a pure data structure. The structure `Direction` contains two numbers representing the PWM values for the motors:

```
struct Direction {int A; int B; };
```

The standard driving directions can thus easily be declared via the associated PWM values for the motors:

```
Direction Forward   = {+255, +255};
Direction Backward  = {-255, -255};
Direction Right     = {+255, -255};
Direction Left      = {-255, +255};
Direction Stop      = {  0,   0};
```

In the class `MotorNew`, the method `drive` is defined, which receives the structure `Direction` as an argument. The new method is created as an inline function that passes the PWM values from the structure `Direction` to the method `write` of the base class `Motor`:

```
class MotorNew : public Motor {
  public:
    void drive (Direction R) { Motor::write(R.A, R.B);};
};
```

7.5 Obstacle Avoidance with two Distance Sensors

The object **motor** of the derived class **MotorNew** is initialized using the standard constructor. If you use a different pin assignment than that of the Arduino motor shield, you would have to provide a further constructor for the derived class, which sets the corresponding pins via the constructor of the base class to be called with four arguments.

```
MotorNew motor;
```

Next, the desired evasive maneuvers are to be defined depending on the return value of the routine **readSensor**. As long as no obstacle is detected, the robot moves straight ahead. If there is an obstacle on only one side (left or right), the robot moves to the other direction. If an obstacle is detected straight ahead, the robot should rotate 180° (see Tab. 7.3).

Table 7.3: Sensor reading and resulting evasive maneuvers

Sensor value	Text	Maneuver
0	None	Straight ahead
1	Left	Evade to the right
2	Right	Evade to the left
3	Straight ahead	180°-turn

The directions intended for the evasive maneuvers are stored in the array **Evasion** according to the numbering of the sensor reading (0...3):

```
Direction Evasion[] = {Forward, Right, Left, Right};
```

The time required for a 180°-turn is to be determined experimentally and stored as value in ms in the constant **T_turn**:

```
// 180 degree turn
const int T_turn = 1000;
```

In the main loop, the sensors are first read and the result of the qualitative evaluation of the measured values is displayed on the LCD. The direction of travel is then set using the method **drive**. In the case of a obstacle detected head-on, the robot carries out a 180°-turn.

```
void loop() {
  int x=readSensor();
  lcd.clear();
  lcd.print(Message[x]);
  motor.drive(Evasion[x]);
  if (x==3) {
    delay(T_turn);
  }
  delay(100);
}
```

7.6 Distance Measurement using Ultrasonic Sensors

7.6.1 Ultrasonic Distance Measurement Principle

In addition to optical triangulation, a number of other methods are available for distance measurement. In mobile robots, distance measurement with ultrasonic sensors is very common. During measurement, an ultrasonic transmitter outputs a signal which is reflected by the object to be measured (obstacle). An ultrasonic microphone (i.e., a receiver) detects the incoming sound wave (see Fig. 7.10).

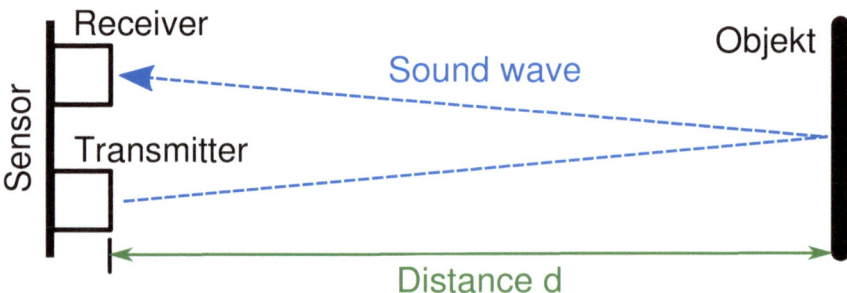

Figure 7.10: Ultrasonic distance measurement method

If the object in question has a distance d from the ultrasonic sensor, the sound must travel twice the distance (i.e., $2d$). If we describe the propagation

7.6 Distance Measurement using Ultrasonic Sensors

time by t and the speed of the sound waves by c, we obtain the following relationship:

$$2d = c \cdot t$$

The distance d therefore is given by the formula:

$$d = t \cdot c/2$$

The speed of sound depends on various factors (e.g. the medium and its temperature) and is approx. $c = 343\,\text{m/s}$ (1.125 ft/s) for air at $20\,°\text{C}$ ($68\,°\text{F}$). In the above formula, half the speed of sound is required. To do this, we define the constant k:

$$k = c/2 = \frac{343}{2}\,\text{m/s} = 171.5\,\text{m/s}$$

If the propagation time t of the sound wave in microseconds (µs) and the distance d in cm are entered, the following applies:

$$k = 171.5\,\text{m/s} = 0.01715\,\text{cm/µs}$$

The distance in cm is therefore approximated by dividing the time measured in µs by 58:

$$k = 0.01715\,\text{cm/µs} \approx \frac{1}{58.3}\,\text{cm/µs}.$$

7.6.2 Operation of the Ultrasonic Distance Sensors HC-SR04 and HY-SRF05

A very popular sensor is the HC-SR04. In addition to the ultrasonic transmitter and receiver, it also has the complete control electronics. The distance sensor is intended for a measuring range from 2 cm to $4\,\text{m} = 400\,\text{cm}$. The HY-SRF05 ultrasonic distance sensor can be seen as a marginal advanced version of the HC-SR04 (see Fig. 7.11).

The HC-SR04 sensor has 4 pins, whereas the HY-SRF05 has 5 pins. In addition to ground (Gnd) and supply voltage (Vcc), two TTL signals are provided for control. The signal Trig starts the measuring process. To do this, set this pin to HIGH level for 10 µs. The propagation time is fed back as a pulse

Figure 7.11: Ultrasonic distance sensors HC-SR04 (left), HY-SRF05 (right)

with HIGH level via pin Echo. In the test setup, the digital connections 0 and 1 of the Arduino board were used for the communication with the ultrasonic distance sensor (Fig. 7.12).

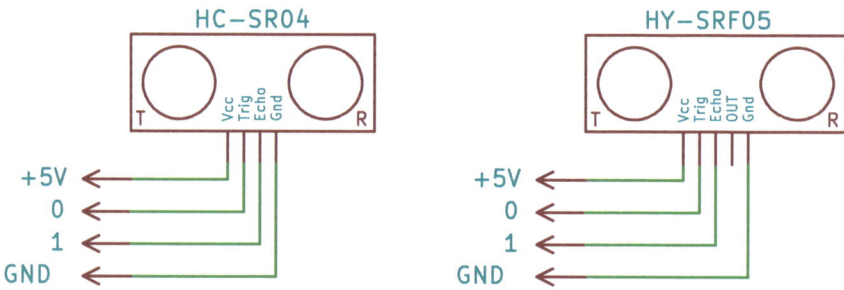

Figure 7.12: Connecting the distance sensor HC-SR04 (left) or HY-SRF05 in compatibility mode (right)

For operation of the HY-SRF05 sensor in compatibility mode (Mode 1) to the HC-SR04, the additional pin remains unconnected. In the following descriptions only the HC-SR04 is considered (which is more common anyway), although the same software can also be used for the HY-SRF05 without any modification.

For distance measurement with the HC-SR04, a suitable Arduino library is available. To do this, download the ZIP file HC_SR4_Demo_Arduino.zip from the company's website [2]. With the menu item Sketch > Include Library > Add .ZIP Library you select the downloaded ZIP file and integrate it into the Arduino IDE. Thereby the subdirectory HC_SR4_Demo_Arduino is created in the Arduino directory libraries, which contains besides the C++ sources (Ultrasonic.h and Ultrasonic.cpp) also an example.

7.6 Distance Measurement using Ultrasonic Sensors

The example program is described below (with small adjustments) [2]. First, the required header files are included:

```
#include <Ultrasonic.h>
#include <LiquidCrystal.h>
```

For the ultrasonic sensor as well as for the LCD, instances of the provided classes must be created. In this case, the example program must be modified with regard to the pin assignment used. The digital channels 0 and 1, which are transferred to the object `ultrasonic` as arguments, are used to query the sensor:

```
LiquidCrystal lcd(10, 2, 7, 6, 5, 4);
Ultrasonic ultrasonic(0,1);
```

In the function `setup` only the LCD has to be initialized in the usual way:

```
void setup() {
  lcd.begin(20, 4);
}
```

The measurement is made in the main loop and the result of the measurement is displayed:

```
void loop() {
  lcd.home();
  lcd.print(ultrasonic.Ranging(CM));
  lcd.print(" cm ");
  delay(100);
}
```

The method `Ranging` of the class `Ultrasonic` returns the distance in cm for the argument `CM` (as in the example), and in inches for the argument `INC`.

The sensor query can also be done without installing the above mentioned additional library [4]. For a good programming style, we define the digital channels used to query the HC-SR04 as constants:

```
const int pinTrig = 0;
const int pinEcho = 1;
```

In the function `setup` the pin used to trigger the measurement is configured as output and the pin used for the signal propagation time is configured as input:

```
void setup() {
  lcd.begin(20, 4);
  pinMode(pinTrig, OUTPUT);
  pinMode(pinEcho, INPUT);
}
```

The function `pulseIn` is available in the Arduino environment to measure a pulse length [1]. The number of the digital channel used must be passed to this function as the first argument. The second argument specifies whether the length of a HIGH or LOW pulse is to be measured (see Fig. 7.13).

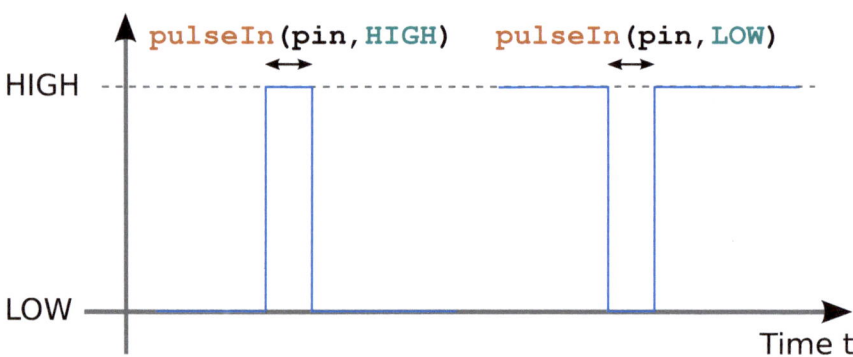

Figure 7.13: Measuring the width of a HIGH or LOW pulse with the Arduino function `pulseIn`

The function `pulseIn` returns the pulse duration (propagation time of the sound wave) in µs as an unsigned long integer number. For this purpose we create the variable `t`:

```
long unsigned t;
```

Additionally, the function `pulseIn` can optionally be given a third argument for aborting the measurement if a certain time is exceeded. When the maximum measuring range of 4 m is reached, the sound must travel 8 m. With the above given sound velocity of approx. 343 m/s this corresponds to a propagation time

7.6 Distance Measurement using Ultrasonic Sensors

of approx. 23.3 ms or 23300 µs. Therefore we specify that the measurement is canceled after 25 ms or 25000 µs:

```
const long unsigned TimeOut = 25000;
```

In the main loop, the pin Trig is first (briefly) set to LOW level. Then a 10 µs pulse with HIGH level is output on this pin, which triggers the measurement process. With `pulseIn` the propagation time is read in. This is followed by conversion and output. Here the distance in mm is specified so that the running time measured with `pulseIn` is to be divided by $5.83 = 583/100$ (instead of 58.3 as for the value in cm).

```
void loop() {
  lcd.home();
  // start measurement
  digitalWrite(pinTrig,LOW);
  delayMicroseconds(2);
  digitalWrite(pinTrig,HIGH);
  delayMicroseconds(10);
  digitalWrite(pinTrig,LOW);
  // read propagation time
  t=pulseIn(pinEcho,HIGH,TimeOut);
  // output
  lcd.print("Value : ");
  lcd.print(t);
  lcd.print("    ");
  lcd.setCursor(0,1);
  lcd.print("Distance : ");
  lcd.print((100*t)/583);
  lcd.print(" mm ");
  delay(100);
}
```

If the runtime defined under `TimeOut` is exceeded, the value zero is returned. This case could also be intercepted by the program and then the maximum allowed distance (i.e., 4 m or 4000 mm) could be displayed.

For the operation of two HC-SR04 sensors, four digital channels would normally be used. In addition to the channels 0 and 1 already used, channels 8 and 9 could be used, if the brake function of the Arduino motor shield is not needed (see Section 3.2). One channel can be saved by combining the Trig connections of both sensors. With the circuit variant shown in Fig. 7.14, the measurement of both sensors is started simultaneously via pin 8 and the left or right value is requested individually via pin 0 or 1, respectively.

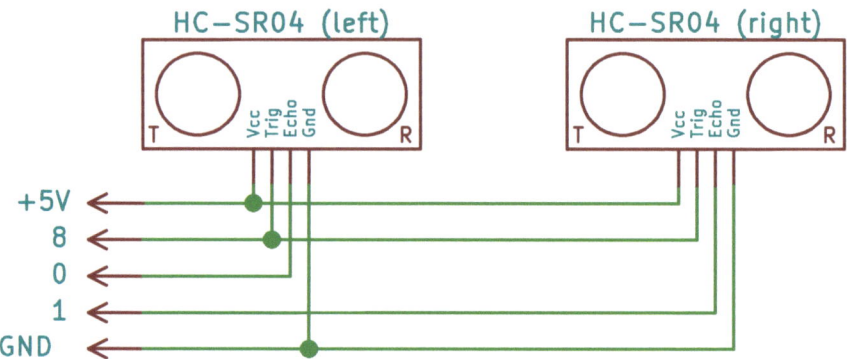

Figure 7.14: Operation of two distance sensors HC-SR04 via three digital channels

7.6.3 Single-wire Operation of the Ultrasonic Distance Sensor HC-SR04

According to datasheet [3], the HY-SRF05 has another operating mode (Mode 2) in addition to the compatibility mode (Mode 1), in which the measurement is triggered and then read in via the same pin. Instead of two digital channels, one channel would be sufficient. Despite intensive tests, the author was not able to perform a distance measurement in this mode. This gave rise to the idea of implementing such an operating mode for the much more popular HC-SR04 sensor.

If you want to transmit both signals (Trig and Echo) over one line, you must prevent two outputs from working against each other. The signal Trig is an input to the sensor, so that we can connect this connector directly to the digital channel of the Arduino board. The pin Echo at the sensor is an output connected to the digital channel of the Arduino board via a 1 kΩ resistor as shown in Fig. 7.15. If the channel on the Arduino board is configured as an output for starting the measurement process, the current flow to the pin Echo is limited by the resistor to $5\,\text{V}/1\,\text{k}\Omega = 5\,\text{mA}$.

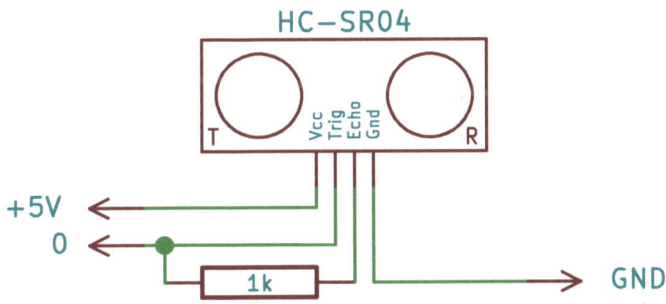

Figure 7.15: Single-wire operation of the ultrasonic distance sensor HC-SR04

Now only one pin is required to operate the sensor:

```
const int pinSensor = 0;
```

In the function **setup** only the LCD has to be initialized. The digital channel for the sensor is configured and reconfigured in the main loop.

```
void setup() {
  lcd.begin(20, 4);
}
```

To trigger the measuring process, the sensor pin is configured as a digital output. The measurement is then started in the usual way by a $10\,\text{µs}$ continuous HIGH pulse. Before reading out the measured value, the sensor connection is reconfigured to a digital input:

```
void loop() {
  lcd.home();
  // configure pin as output
  pinMode(pinSensor, OUTPUT);
  // start measuring
  digitalWrite(pinSensor, LOW);
  delayMicroseconds(2);
  digitalWrite(pinSensor, HIGH);
  delayMicroseconds(10);
  digitalWrite(pinSensor, LOW);
  // reconfigure pin as input
```

```
  pinMode(pinSensor, INPUT);
  // read propagation time
  t=pulseIn(pinSensor,HIGH,TimeOut);
  // output
  lcd.print("Value : ");
  lcd.print(t);
  lcd.print("    ");
  lcd.setCursor(0,1);
  lcd.print("Distance : ");
  lcd.print((100*t)/583);
  lcd.print(" mm ");
  delay(100);
}
```

For the operation of two distance sensors, which are typically mounted on the left and right side of the robot's front, only two digital channels are now required. Fig. 7.16 shows the associated circuit diagram.

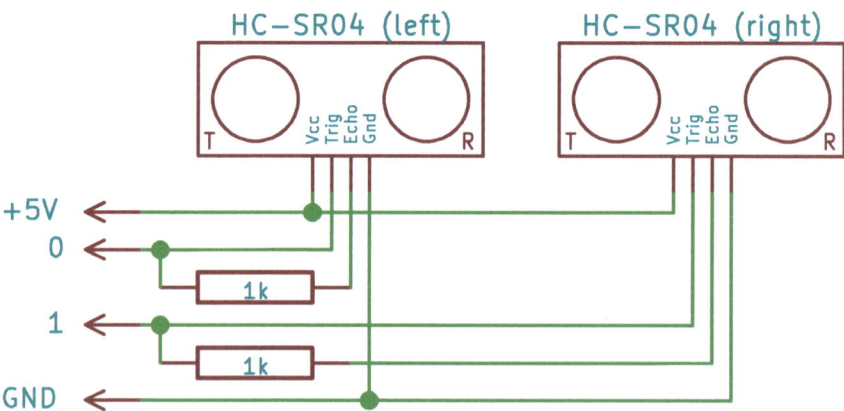

Figure 7.16: Single-wire operation of two ultrasonic distance sensors HC-SR04

References

[1] *Arduino — Home.* `https://arduino.cc`.

[2] *Cytron Technologies – Ultrasonic Ranging Module.* `https://www.cytron.com.my/p-sn-hc-sr04`.

[3] *SRF05 Technical Documentation.* `http://www.robot-electronics.co.uk/htm/srf05tech.htm`.

[4] *Ultrasonic Sensor HC-SR04 and Arduino Tutorial.* `https://howtomechatronics.com/tutorials/arduino/ultrasonic-sensor-hc-sr04/`.

[5] Sharp: *GP2Y0A02YK0F, Distance Measuring Sensor Unit, Measuring distance: 20 to 150 cm, Analog output type.* Datasheet.

[6] Sharp: *GP2Y0A21YK0F, Distance Measuring Sensor Unit, Measuring distance: 10 to 80 cm, Analog output type.* Datasheet.

[7] Sharp: *GP2Y0A41SK0F, Distance Measuring Sensor Unit, Measuring distance: 4 to 30 cm, Analog output type.* Datasheet.

Chapter 8

Line Detection and Tracking

8.1 Line Detection with a Reflective Optical Sensor

Playing field and road markings can be detected with a reflective optical sensor. An LED emits a light beam whose reflection is detected by a phototransistor (Fig. 8.1). Infrared light can be used to suppress the influence of daylight.

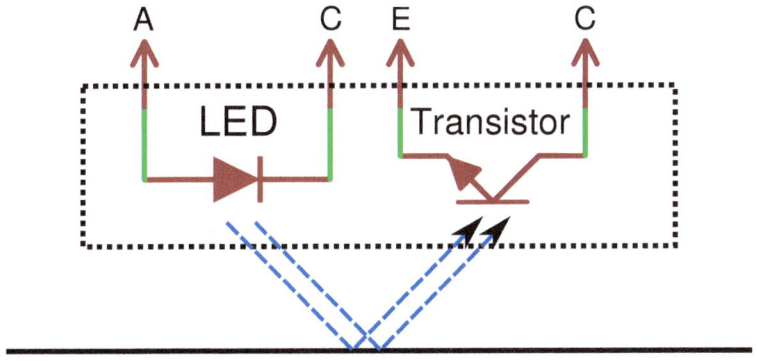

Figure 8.1: Principle of a reflective optical sensor

The integrated reflective optical sensor CNY70 [1] is very common. LED and phototransistor are mounted in a 4-pin package (see Fig. 8.2).

Figure 8.2: Two reflective optical sensors CNY70

The LED is connected via a 560 Ω series resistor to the supply voltage of +5 V, the phototransistor via a 47 kΩ resistor (Fig. 8.3). The collector voltage of the phototransistor is connected to an analog input (here: A4) of the Arduino board. With a white background a lot of light is reflected, so that the transistor conducts. Thus a low voltage is applied to A4. On a black background, on the other hand, hardly any light is reflected. In this case the transistor is blocked and almost the full supply voltage is applied to A4.

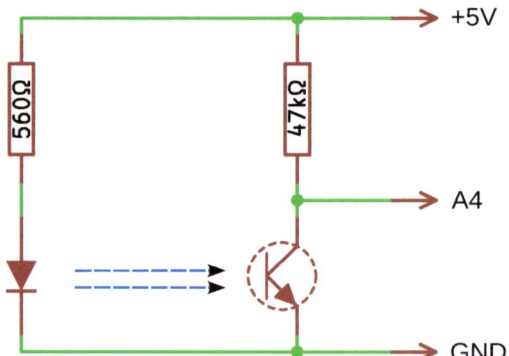

Figure 8.3: Wiring of the reflective optical sensor

Fig. 8.4 shows the wiring of the reflex optical sensor CNY70 taking the pin assignment into account, see [1]. Constructively the reflex optical sensor should have a maximum distance of about 5 mm (i.e., 0.2 in) from the playing field or the ground (cf. Fig. 8.5).

8.1 Line Detection with a Reflective Optical Sensor

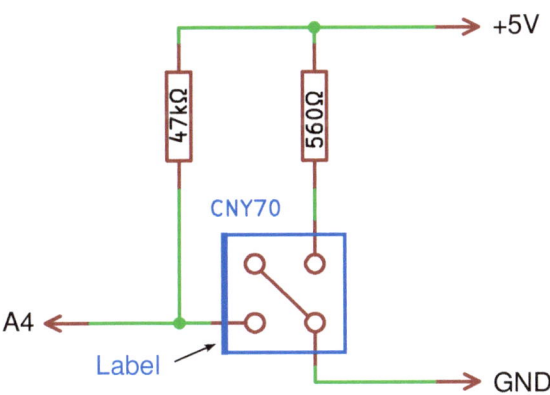

Figure 8.4: Wiring of the reflective optical sensor CNY70 (view from the connector side, label on the left side)

Figure 8.5: Line sensor with CNY70 on the front side of the robot

Sensor reading is practically the same as with the Sharp distance sensor. We assume that the LCD has been integrated into the software in the usual way. The analog input used for the sensor is stored as constant `pinLine`, for measurement result and output the variables `x` and `line` are declared:

```
const byte pinLine = A4;
unsigned x=0;
char line[21];
```

In the main loop, the analog input is read (without filtering) and displayed on the LCD panel:

```
void loop() {
  x = analogRead(pinLine);
  lcd.home();
  sprintf(line, "ADC: %4u", x);
  lcd.print(line);
  delay(200);
}
```

For white background, ADC values of a maximum of 100 were read in, for black background of approx. 800 and more. However, these values can differ depending on the used resistance values or the distance of the reflective optical sensor from the ground.

8.2 Line Tracking with one Reflective Optical Sensor

Line tracking control can be implemented with the line sensor. This section introduces a simple implementation. To describe the submaneuvers that make up the line following control, we reintroduce the structure `Direction`, which contains the PWM values for the motor control:

```
struct Direction {int A; int B; };
```

Based on this structure we define the required directions. Forward movement should not be at full speed. With `Slight_Left` or `Slight_Right`, one motor rotates in forward direction, the other stops. Thus one has on the one hand a

8.2 Line Tracking with one Reflective Optical Sensor

turn into the corresponding direction, on the other hand also a slight forward movement.

```
Direction Forward    =   {200, 200};
Direction Slight_Right = {200,   0};
Direction Slight_Left  = {  0, 200};
```

To hand over the direction defined as a structure to the motors, we derive the class `MotorNew` from the basic class `Motor` as in the last chapter and create the method `drive`. Of course, the header file `Motor.h` must be included first.

```
class MotorNew : public Motor {
  public:
    void drive (Direction R) { Motor::write(R.A, R.B);};
};
```

The number of the analog input channel used for the CNY70 is stored in the next constant:

```
const byte pinLine = A4;
```

The robot should drive straight ahead until it detects a line. After that, let the robot follow this line. The variable `track` is used to distinguish between these two states — search for the line or line tracking. This variable is first assigned the value `false`; if a line is detected, the value is set to `true`:

```
bool track = false;
```

On a white or light background, the robot moves either straight ahead (if the lane has not yet been detected) or slightly to the right (to return to the black line). After recognizing the black line, the robot moves along the white-black edge, so to speak.

```
void loop()
{
  // black line found?
  if (analogRead(pinLine) > 500) {
    motor.drive(Slight_Left);
    track = true;
  }
  else {
    if (track)
```

```
      motor.drive(Slight_Right);
    else
      motor.drive(Forward);
  }
}
```

Of course, the program can still be augmented by a suitable LCD output.

8.3 Line Tracking with Several Reflective Optical Sensors

Line tracking control can be improved using more than one line sensor. It makes sense to use two or three line sensors. Fig. 8.6 shows a mobile robot with three line sensors.

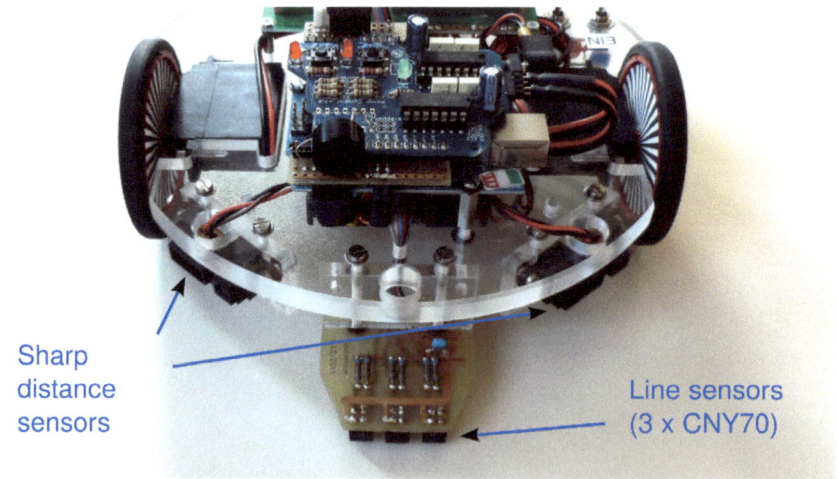

Figure 8.6: Mobile robot with two distance and three line sensors

If you want to use several line sensors together with several distance sensors, the six analog inputs of the standard boards may no longer be sufficient. The following solutions are conceivable here:

- The motor current measurement signals of the Arduino motor shield are not used. By cutting the corresponding solder bridges on the conductor

8.3 Line Tracking with Several Reflective Optical Sensors

side of the motor shield, the inputs A0 and A1 could be used for the sensors (see Fig. 3.6 on p. 21).

- You use an Arduino board with more analog inputs, e.g. Arduino Mega or Arduino Mega 2560 (see Section 1.2.3).

- A board is used in which digital channels can also be utilized as analog channels. This applies to the boards Arduino Leonardo and Arduino Micro (see Sections 1.2.2 and 1.2.3).

With three sensors, one would no longer have to orientate oneself on the white-black or black-white edge, but could also drive directly on the line. In the program, you would first declare the analog channels used as constants:

```
// analog input pins for line sensors
const byte pinLineL = A3; // left
const byte pinLineC = A4; // center
const byte pinLineR = A5; // right
```

The qualitative evaluation of measured values can be implemented via a single function:

```
byte LineSensor() {
  const int MAX=500;
  byte x;
  x = analogRead(pinLineL)<MAX ? 0 : 1;
  x+= analogRead(pinLineC)<MAX ? 0 : 2;
  x+= analogRead(pinLineR)<MAX ? 0 : 4;
  return x;
}
```

Tab. 8.1 classifies the possible sensor states and gives suggestions for the submaneuvers to be implemented. For example, if the robot travels exactly on the line, i.e., either only the middle sensor or all three sensors detect black (return value 2 or 7 of the line sensor function), the robot can continue to travel straight ahead quickly.

Table 8.1: States of the three line sensors

Value	Sensor left	Sensor center	Sensor right	Submaneuver
0	white	white	white	Search_Line
1	black	white	white	Strong_Left
2	white	black	white	Fast_Forward
3	black	black	white	Slight_Left
4	white	white	black	Strong_Right
5	black	white	black	Crossroad
6	white	black	black	Slight_Right
7	black	black	black	Fast_Forward

These suggestions still have to be translated into an executable program. There are also numerous possibilities to incorporate your own ideas.

References

[1] Vishay Telefunken: *CNY70, Reflective Optical Sensor with Transistor Output*. Datasheet.

Chapter 9

Wireless Control of the Robot

9.1 Infrared Remote Control

Infrared remote controls can be found in almost every household, for example for the television and the audio system. Many small models (e.g. miniature helicopters for indoor use) are also controlled by infrared (IR). Therefore, it makes sense to use an already existing IR remote control to operate the mobile robot. These remote controls emit a modulated signal via an IR LED whose carrier frequency is typically in the range of 30...56 kHz.

9.1.1 Connecting the Receiver Module and Requesting the IR Codes

In principle, an IR photodiode or an IR phototransistor could be used to receive an IR signal. However, it is better to use an integrated IR receiver module such as the PNA4602 or the TSOP4838 (Fig. 9.1), where the received signal is amplified and preprocessed. The mentioned receiver moduls are designed for the very common carrier frequency of 38 kHz and should therefore work with most remote controls. Numerous similar IR receivers are available from specialized retailers.

Figure 9.1: IR receiver module TSOP4838

To connect the TSOP4838 IR receiver, a digital input is required in addition to the supply voltage and ground from the Arduino board. In Fig. 9.2 the digital channel 0 was provided for this. However, some IR receivers have a different pin assignment.

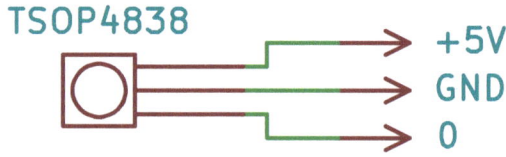

Figure 9.2: Wiring of the IR receiver module TSOP4838

The evaluation of the signal supplied by the IR receiver module is done with the help of the library **Arduino-IRremote**. (This library also supports the control of an IR transmitter, so that you can build a universal remote control with Arduino [4].) To install the library, proceed as follows:

1. Download the library **Arduino-IRremote** as ZIP file, e.g. via GitHub [2].

2. To add the library to the Arduino development environment, either select the ZIP file via the menu item Sketch > Include Library > Add .ZIP Library or unpack the ZIP file directly into the Arduino directory **libraries**.

3. Remove the subdirectory **RobotIRremote** from the Arduino program directory **libraries**, which is typically located in **/usr/share/arduino** under Linux.

9.1 Infrared Remote Control

The third step is necessary because the files `IRremote.h` and `IRremote.cpp` conflict with the files of the same name in the library `RobotIRremote` of the less common official Arduino robot [1]. The library `Arduino-IRremote` should then be available (after a restart of the Arduino IDE if necessary). Under the menu item File > Examples > IRremote various sample programs are stored for this library. By calling the example program `IRrecvDemo` you can test the IR receiver, whereby only the used digital pin of the Arduino board has to be adapted. (In the example program pin 11 is used, which is already occupied by the Arduino-Motor-Shield.) When an IR remote control is activated, the received control code is displayed via the serial monitor of the Arduino IDE (Tools > Serial Monitor).

Based on the example program of the library, the IR codes are to be displayed in a similar way on the robot's LCD panel. The first step is to integrate the required libraries:

```
#include <LiquidCrystal.h>
#include <IRremote.h>
```

The display is treated as usual:

```
LiquidCrystal lcd(10, 2, 7, 6, 5, 4);
```

When initializing the object responsible for decoding the IR signal, the digital pin used for the receiver must be provided:

```
const int pinIR = 0;
IRrecv irrecv(pinIR);
```

To store the values obtained during decoding, create another object:

```
decode_results results;
```

In the function `setup`, the LCD is initialized and the query of the IR receiver is started:

```
void setup() {
  lcd.begin(20, 4);
  irrecv.enableIRIn();
}
```

The main loop displays any decoded signal on the LCD panel:

```
void loop() {
  if (irrecv.decode(&results)) {
    irrecv.resume();
    lcd.clear();
    lcd.print("IR code: ");
    lcd.print(results.value, HEX);
  }
}
```

9.1.2 Motor Control via IR Codes

The example program described for displaying the IR codes is to be extended to include motion control of the robot. For the control of the motor we integrate the library `Motor`. Additionally, we assume that the class `MotorNew`, which provides the methods `forward`, `backward`, `right`, `left` and `stop`, was derived from the base class `Motor` as described in Section 7.4 and instantiated accordingly:

```
MotorNew motor;
```

To control the robot, a total of five buttons for forward and reverse, left and right rotation and stop must be selected on an existing remote control. The control codes emitted by the remote control when the corresponding buttons are pressed can be displayed with the program described in the last section and must now be stored as constants. The following codes have been determined for the remote control used in the test:

```
// IR codes of the remote control
const long IR_forward  = 0xB4B4E21DL;
const long IR_backward = 0xB4B412EDL;
const long IR_left     = 0xB4B49A65L;
const long IR_right    = 0xB4B45AA5L;
const long IR_stop     = 0xB4B41AE5L;
```

The variable `IRcode` is declared for the current query result:

```
long IRcode;
```

9.1 Infrared Remote Control

In the main loop, the IR code is first displayed on the LCD. If the robot matches one of the stored IR codes, the corresponding movement of the robot is initiated:

```
void loop() {
  // query IR code
  if (irrecv.decode(&results)) {
    IRcode = results.value;
    irrecv.resume();
    lcd.clear();
    lcd.print("IR code: ");
    lcd.print(IRcode, HEX);
    // display and motor control
    lcd.setCursor(0,1);
    switch(IRcode) {
      case IR_forward:
        lcd.print("Forward");
        motor.forward();
        break;
      case IR_backward:
        lcd.print("Backward");
        motor.backward();
        break;
      case IR_left:
        lcd.print("Left");
        motor.left();
        break;
      case IR_right:
        lcd.print("Right");
        motor.right();
        break;
      case IR_stop:
        lcd.print("Stop");
        motor.stop();
        break;
      default:
        lcd.print("Code unknown");
    }
  }
}
```

As an extension of the described test program, for example, combined movements (e.g. right forward) can be provided or the driving speed can be set via two further buttons on the remote control (faster, slower).

9.2 Radio Remote Control

9.2.1 Connecting the Receiver

Radio remote controls are used in car, ship and aircraft model construction, among others. Various frequency bands are available for operating remote controls, e.g. 27 MHz, 35 MHz or 2.4 GHz. Here, the HT-4 transmitter was used in combination with the HR-4 receiver from Reely [3], see Fig. 9.3. In fact almost any other remote control can be used. It is only important that the transmitter and receiver communicate correctly with each other.

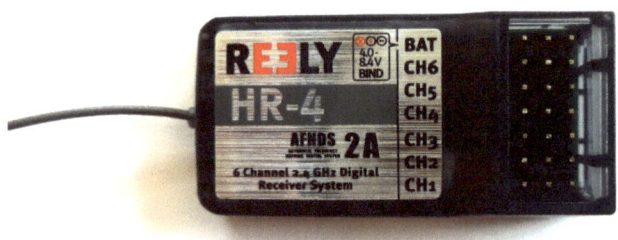

Figure 9.3: Receiver HR-4 from Reely

The usual remote controls are primarily designed for the control of *servos* [6]. Servos are connected to the remote control receiver via three lines (supply voltage, ground, signal). Even though there are several different variants of connection, the common servos follow the wiring diagram shown in Fig. 9.4 with the middle pin for the supply voltage. This scheme is also used for motor control of RC models.

9.2 Radio Remote Control

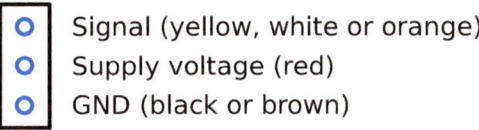

Figure 9.4: Typical wiring of servos

The remote control used is a 4-channel system, even if the design of the receiver with connectors CH1 to CH6 suggests a 6-channel system. The abbreviation CH stands for *channel*. Only two channels are required for the robot. We use the channels CH1 and CH2, which are typically occupied by the aileron and elevator servo on aircraft models. The servo outputs are connected to the digital channels 0 and 1 of the Arduino board (see Fig. 9.5). The receiver is connected to ground and supply voltage via the additional connector BAT. This general connection scheme can also be found in numerous other receivers.

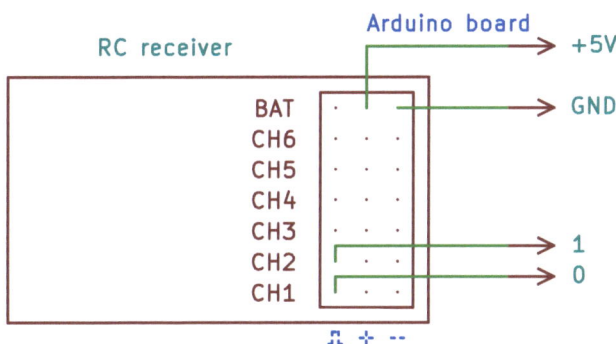

Figure 9.5: Wiring of the receivers HR-4

The servos are controlled via pulse width modulation (PWM). The middle position of the servo is set via a HIGH pulse with a duration of 1.5 ms. For the left or right stop, the servo expects HIGH pulses with a duration of 1 ms or 2 ms, respectively. The relevant pulses are generated with a frequency of approx. 50 Hz, i.e., in a time grid of approx. 20 ms (see [5] and Fig. 9.6).

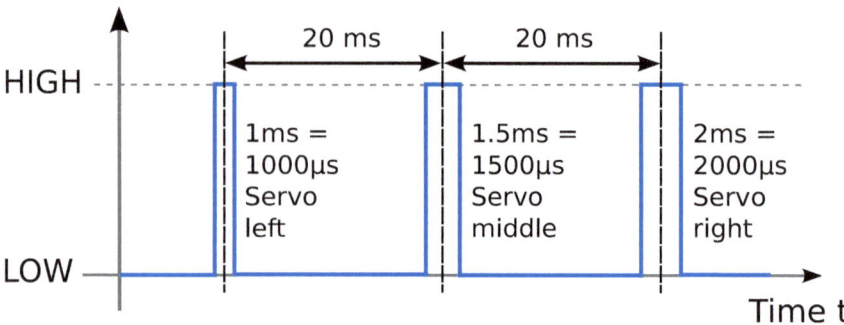

Figure 9.6: Servo signals (PWM)

For an initial test of the remote control, the LCD module is integrated in the usual way:

```
#include <LiquidCrystal.h>
LiquidCrystal lcd(10, 2, 7, 6, 5, 4);
```

The digital channels 0 and 1 used for reading the servo signals are declared as constants:

```
// pins for receiver
const int pinCH1 = 0;
const int pinCH2 = 1;
```

In the function **setup** you set the size of the used LCD panel as usual. The pins used for remote control are declared as inputs:

```
void setup()
{
  lcd.begin(20, 4);
  pinMode(pinCH1, INPUT);
  pinMode(pinCH2, INPUT);
}
```

The pulse lengths of the servo signals are read in the main loop with the function **pulseIn** and displayed on the LCD panel:

```
void loop() {
  // read receiver
  long unsigned x = pulseIn(pinCH1,HIGH);
  long unsigned y = pulseIn(pinCH2,HIGH);
  // output
  lcd.home();
  lcd.print("CH 1: ");
  lcd.print(x);
  lcd.print("    ");
  lcd.setCursor(0,1);
  lcd.print("CH 2: ");
  lcd.print(y);
  lcd.print("    ");
}
```

If the remote control works correctly, the pulse lengths from 1000 µs to 2000 µs of the two remote control channels should also display numerical values in the range from approx. 1000 to 2000.

9.2.2 Conversion of the Signals for Motor Control

In the last section the connection between the receiver of the remote control and the Arduino board was established on both the hardware and software sides. The obtained signals must be adapted for the motor control. This section deals with the required transformation. It is assumed that the motor and, if required, the LCD module are integrated on the software side in the usual way.

The remote control receiver should provide PWM signals with pulse widths in the range of 1000 µs to 2000 µs. Actual values may vary slightly depending on the remote control. With the program presented in the previous section, we can not only check the function of the remote control, but also read the measured pulse widths. With the remote control used, pulse lengths in the range from 1300 to 2500 (both in µs) were measured. We store the limit values (minimum and maximum values) in two constants:

```
// pulse widths in ms
const int TMIN = 1300;
const int TMAX = 2500;
```

In the main loop, you would first insert the code already introduced in the previous section for querying the pulse widths:

```
// get pulse widths
long unsigned x = pulseIn(pinCH1,HIGH);
long unsigned y = pulseIn(pinCH2,HIGH);
```

If required, the obtained values can of course be displayed on the LCD. The signal values would be restricted to the value range described by the constants `TMIN` and `TMAX` using the function `constrain`:

```
// limit pulse widths
int u = constrain(x,TMIN,TMAX);
int v = constrain(y,TMIN,TMAX);
```

In the next step, the range of `TMIN` and `TMAX` is scaled to the range of the motor signals (typically -255 to $+255$) using the function `map`:

```
// scaling
u = map(u,TMIN,TMAX,-255,+255);
v = map(v,TMIN,TMAX,-255,+255);
```

To test the program created so far, the calculated values could be passed directly to the motors:

```
motor.write(u,v);
```

The channels 1 and 2 (CH1 and CH2) of the remote control are often linked to two directions of a control stick. With this simple control method, both motors should rest in the middle position of the control stick. If the stick is at the top right, the robot moves forward, if the stick is in the bottom left position it moves backwards, see Fig. 9.7 (left). If the control stick is at the top left or bottom right, then the robot rotates to the left or to the right. If a different behaviour occurs, the orientation (left/right) of the respective channel may have to be changed. Many remote controls are equipped with additional switches for this purpose. You can also change the orientation on the software by modifying the call `map(u,TMIN,TMAX,-255,+255)` to `map(u,TMIN,TMAX,+255,-255)` for the variable u (change of sign).

9.2 Radio Remote Control

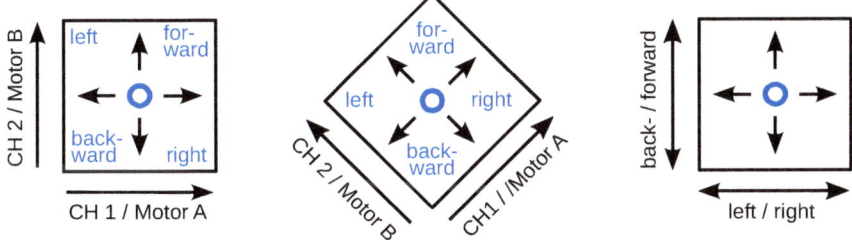

Figure 9.7: Mapping of a remote control stick to the motor signals, simple variant (left), remote control rotated by 45° (centre), corrected variant (right)

The initially implemented simple control of the motors is somewhat unusual. Turning the remote control transmitter counter-clockwise by 45° results in a much more intuitive operation, see Fig. 9.7 (middle). A counter-clockwise rotation (in mathematically positive sense of rotation) with the angle α from a position (x, y) to the new position (x', y') can be described by the following formula:

$$\begin{aligned} x' &= x \cos(\alpha) + y \sin(\alpha) \\ y' &= -x \sin(\alpha) + y \cos(\alpha) \end{aligned}$$

The trigonometric functions sine and cosine have the following functional value at an angle of 45°:

$$\sin(45°) = \cos(45°) = \frac{1}{2}\sqrt{2} \approx 0,7071 \approx 0,7.$$

In our program, the position described by (u, v) is transformed into the new variables (a, b). The fractional rational number $0.7 = 7/10$ is implemented as a decimal fraction:

```
// 45 degree rotation
int a = ( 7*u + 7*v) / 10;
int b = (-7*u + 7*v) / 10;
```

The variables u and v are limited to the range from -255 to +255. With the conversion into variables a and b, this range can be exceeded by up to 40%. This is no problem with the class Motor presented in this book, as it is automatically limited to the permissible range of ±255. If another library is used for motor control, a suitable limitation may have to be made, e.g. with the

Arduino function `constrain`:

```
// limit motor signals
a = constrain(a,-255,255);
b = constrain(b,-255,255);
```

Instead of variables u and v, we now use variables a and b for motor control:

```
// motor control
motor.write(a,b);
```

With this conversion you would also move forward with the forward position of the control stick, backward with the reverse position and so on as shown in Fig. 9.7 (right).

Up to now it was implicitly assumed that the remote control receiver would output the corresponding pulses. This is not the case, for example, when the transmitter is switched off or when the receiver is disconnected from the Arduino board. To recognize this situation, we define the constant `TIMEOUT` in such a way that the pulse measurement is aborted after 50 ms. This is more than twice the PWM cycle time, see Fig. 9.6 on p. 120. The constant `TMID` specifies the mean pulse duration, which corresponds to the center position of the control stick:

```
// timeout und mean value
const unsigned long TIMEOUT = 50000;
const int TMID = (TMIN + TMAX)/2;
```

In the main loop, the function `pulseIn` would be called with an additional third argument (see Section 7.6.2):

```
// get pulse widths
long unsigned x = pulseIn(pinCH1,HIGH,TIMEOUT);
long unsigned y = pulseIn(pinCH2,HIGH,TIMEOUT);
```

If no HIGH pulse is detected within the time specified by the constant `TIMEOUT`, the function `pulseIn` returns the value zero. In this case, we set the variables x and y to the average pulse duration:

```
// set mean value on timeout
if (x==0) x = TMID;
if (y==0) y = TMID;
```

Both motors would stop if the remote control is switched off.

References

[1] *Arduino — Robot.* `https://www.arduino.cc/en/Main/Robot`.

[2] *Arduino-IRremote. Infrared remote library for Arduino: send and receive infrared signals with multiple protocols.* `https://github.com/z3t0/Arduino-IRremote`.

[3] Reely: *Remote Control „HT-4" 2.4 GHz.* User Manual.

[4] Schmidt, M.: *Arduino: A Quick-Start Guide.* The Pragmatic Programmers, 2nd edition, 2015.

[5] Wikipedia contributors: *Servo control — Wikipedia, the free encyclopedia.* `https://en.wikipedia.org/w/index.php?title=Servo_control`.

[6] Wikipedia contributors: *Servo (radio control) — Wikipedia, the free encyclopedia.* `https://en.wikipedia.org/w/index.php?title=Servo_(radio_control)`.

Chapter 10

Additional Design Variants

10.1 LCD KeyPad Shield

If you want to avoid the wiring of an LCD module with the Arduino board as described in Chapter 4, you can use an LCD KeyPad Shield, which is provided by different suppliers. A common variant is the LCD KeyPad Shield from DFRobot [1] shown in Fig. 10.1. This shield has a 16×2 LCD module with backlight, 5 function keys and an additional reset button.

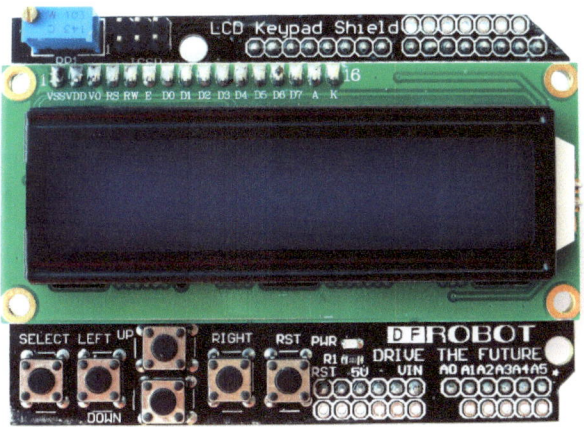

Figure 10.1: LCD KeyPad Shield

The pin assignment of the LCD keypad shield is fixed (hard-wired) and listed in Tab. 10.1. It does not correspond to the pin assignment of the LCD module used in Chapter 4.

Table 10.1: Pin assignment of the LCD KeyPad Shields

Channels (Arduino)	Signals on LCD Modul
4	D4
5	D5
6	D6
7	D7
8	RS
9	E (Enable)
10	Backlight (!)
A0	Buttons

The LCD library must be included in the usual way:

```
#include <LiquidCrystal.h>
```

During instantiation, the object `lcd` must be called with the addresses specified in Tab 10.1:

```
LiquidCrystal lcd(8, 9, 4, 5, 6, 7);
```

The LCD module is initialized in the function `setup`. To do this, the number of rows and columns must be passed:

```
void setup() {
  lcd.begin(16, 2);
}
```

Pin 10 can be used to switch the backlight on or off. If the backlight should always be active, then this pin does not have to be configured, as a suitable pull-up resistor is already provided on the shield.

The shield has 5 buttons wired to the analog channel A0 as shown in Fig. 10.2. The following program performs a simple reading of the associated analog channel:

10.1 LCD KeyPad Shield

```
void loop() {
  int a = analogRead(A0);
  lcd.home();
  lcd.print("LCD KeyPad Shield");
  lcd.setCursor(0,1);
  lcd.print(a);
  lcd.print("    ");
}
```

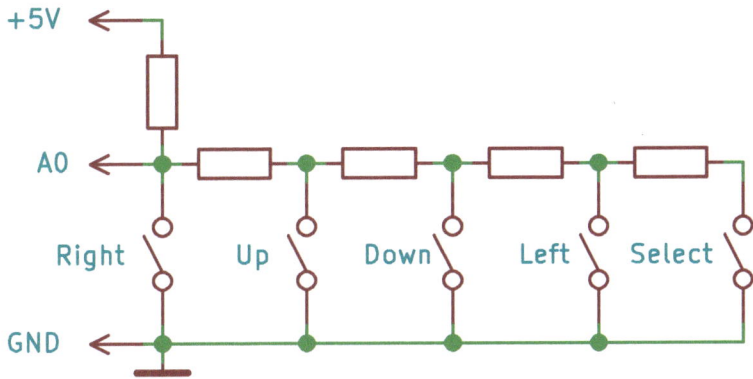

Figure 10.2: Wiring of the push-buttons on the LCD KeyPad Shield

Table 10.2: Values read from the keys of the LCD KeyPad Shield

Push-button	ADC value	Return value
Right	0	0
Up	100	1
Down	256	2
Left	403	3
Select	638	4
No key pressed	1023	5

The corresponding output values of the module used can be found in Tab. 10.2. The values 0 to 4 are provided for the different buttons. If no key is pressed, the value 5 is returned:

```
byte readKey() {
  int a = analogRead(A0);
  if (a>830) return 5;
  if (a>520) return 4;
  if (a>330) return 3;
  if (a>178) return 2;
  if (a> 50) return 1;
  return 0;
}
```

The following main program can be used to validate the key query:

```
void loop() {
  byte a = readKey();
  lcd.home();
  lcd.print("LCD KeyPad Shield");
  lcd.setCursor(0,1);
  lcd.print("Button: ");
  lcd.print(a);
}
```

Operation with the Arduino Motor Shield R3: The signals RS and E of the LCD KeyPad Shield occupy the same digital channels as the brake signals (Brake A/B) of the motor shield. The same problem occurs with the analog channel A0, which is provided for key query on the LCD keypad shield and for current measurement of channel A on the motor shield. The relevant signals are soldered on the back of the motor shield and can be easily disconnected (see Fig. 3.6 on p. 21). With this measure both shields can be operated together.

Operation with the Velleman Motor Shield: Signals PWMA and PWMB could be assigmened to the default channels 3 and 9. Since digital channels 4 and 7 are occupied by the LCD keypad shield, only pin 2 can be used for DIRA. In a similar way, only pin 11 is available for DIRB. The pin assignment is shown in Fig. 10.3.

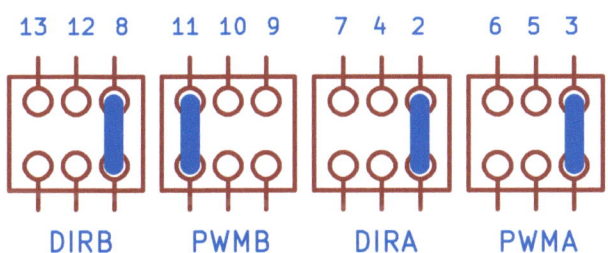

Figure 10.3: Address selection of the Velleman Motor Shield for operation with the LCD KeyPad Shield

10.2 Motor Driver with L298N

In addition to dedicated motor drivers for the Arduino environment (the so-called *motor shields*), there are also motor control modules that are intended for any microcontroller platform. Fig. 10.4 shows such a motor driver. Such modules are also available in similar designs. The heart of the circuit board is a driver circuit of the L298 family. In contrast to the SMD version L298P used for both motor shields (Arduino motor shield and Velleman motor shield), the L298N used here is in a Multiwatt package [2]. The supply voltage input used for motor control is marked by +12 V. The driver circuit L298 is designed for an operating voltage of max. 46 V, but the driver module only up to 35 V. With an operating voltage of up to 12 V, the module can also provide a +5 V auxiliary voltage, which can be used, for example, to supply TTL logic circuits. If the supply voltage for the motors is higher than 12 V, the corresponding jumper must be removed.

In the depicted motor driver module, the pins ENA and ENB (Enable A/B) are equipped with jumpers. For speed adjustment, these two jumpers must be removed and the connections ENA and ENB must be driven with PWM signals. The direction of rotation of one channel (motor A) is set via signals IN1 and IN2, and that of the other channel (motor B) via IN3 and IN4 (see Tab. 10.3). For the motor to rotate, the signals IN1, IN2 or IN3, IN4 must have different levels. At the same level, the motors are either freewheeling (Enable on LOW) or actively braked (Enable on HIGH).

Figure 10.4: Motor driver module with L298N

Table 10.3: Signals of the L298N motor driver module

Signals	Motor A	Motor B
Pulse width modulation (PWM)	ENA	ENB
Direction (DIR)	IN1	IN3
Complementary direction	IN2	IN4

10.2 Motor Driver with L298N

Fig. 10.5 shows the wiring of the motor driver module. The pin assignment used corresponds as far as possible to that of the Arduino motor shield. The digital channels otherwise provided for the brake function were used for the complementary direction signals.

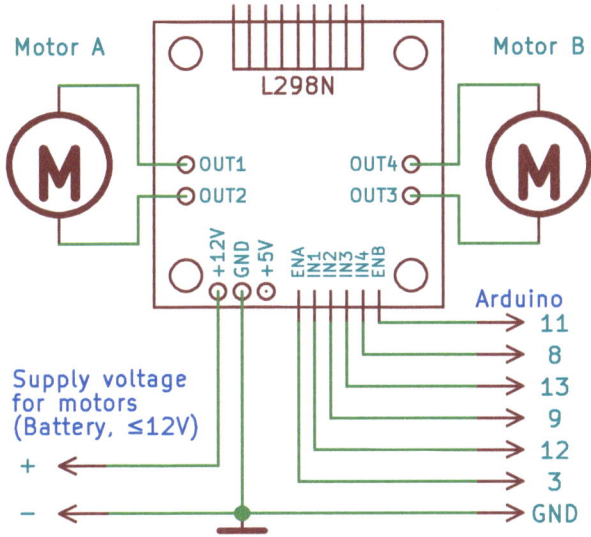

Figure 10.5: Wiring of the L298N motor driver module

```
const byte PWMA  =   3;
const byte PWMB  =  11;
const byte DIRA  =  12; // directions
const byte DIRB  =  13;
const byte DIRA2 =   9; // complementary directions
const byte DIRB2 =   8;
```

In the function setup, these 6 digital channels are defined as outputs:

```
void setup()
{
  pinMode(PWMA,   OUTPUT);
  pinMode(PWMB,   OUTPUT);
  pinMode(DIRA,   OUTPUT);
  pinMode(DIRB,   OUTPUT);
  pinMode(DIRA2,  OUTPUT);
  pinMode(DIRB2,  OUTPUT);
}
```

In the main loop, driving the motors at half speed could look as follows:

```
void loop()
{
  // direction
  digitalWrite (DIRA,  LOW);
  digitalWrite (DIRA2, HIGH);
  digitalWrite (DIRB,  LOW);
  digitalWrite (DIRB2, HIGH);
  // half speed
  analogWrite (PWMA, 127);
  analogWrite (PWMB, 127);
}
```

10.3 Obstacle Detection with IR Sensors

In many cases, one is not interested in precise distance information, but simply wants to detect obstacles in time to initiate a suitable evasive maneuver. For such a request the sensor module FC-51 shown in Fig. 10.6 can be used.

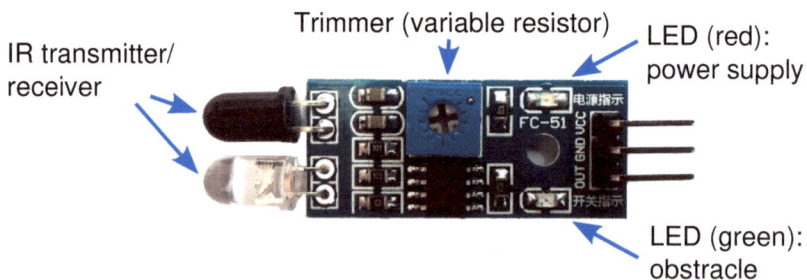

Figure 10.6: IR obstacle detaction sensor modul FC-51

The sensor module has three pins: Ground (GND), supply voltage (VCC) and a digital signal output (OUT). If the power supply is correctly connected, the red LED on the sensor module lights up. An obstacle is indicated by a LOW level at the output. In addition, the green LED on the sensor board lights up. The desired distance from which an obstacle is detected can be adjusted by the trimmer (i.e., a variable resistor) on the PCB.

10.3 Obstacle Detection with IR Sensors

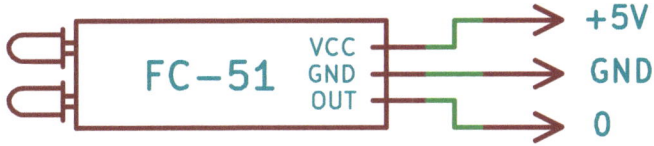

Figure 10.7: Wiring of the IR sensor modul FC-51

Usually the IR sensor will be placed in the middle of the mobile robot's front. To avoid an obstacle, only slightly modify the program already specified for use with a Sharp distance sensor (see Section 7.3). The pin used for the IR sensor is defined by the constant `pinIR` (here: 0). We also create the variable `sensor` for the signal value:

```
// pin IR sensor
const byte pinIR = 0;
byte sensor;
```

The main loop is modified marginally with regard to the sensor query. With a HIGH level at the sensor output the robot moves straight ahead, at an obstacle indicated by LOW level the evasive maneuver is initiated. We use the constants `T_back` and `T_rot` as defined in Sections 5.2, 6.4 and 7.3.

```
void loop() {
  sensor = digitalRead(pinIR);
  if (sensor == HIGH) {
    // forward
    motor.forward(); }
  else {
    // backward
    motor.backward();
    delay(T_back);
    // rotate
    motor.right();
    delay(T_rot); }
  delay(100);
}
```

References

[1] *DFRobot.* https://www.dfrobot.com/.

[2] STMicroelectronics: *L298; Dual Full-Bridge Driver*, 2000. Datasheet.

www.ingramcontent.com/pod-product-compliance
Lightning Source LLC
Chambersburg PA
CBHW040217220526
45473CB00001B/21